The Musculoskeletal System
Physiological Basics

The Musculoskeletal System
Physiological Basics

James Gibson Gamble, M.D., Ph.D.

Division of Orthopaedic Surgery
Stanford University School of Medicine
Children's Hospital at Stanford
Palo Alto, California

Raven Press ◆ New York

Raven Press, 1185 Avenue of the Americas, New York, New York 10036

© 1988 by Raven Press, Ltd. All rights
reserved. This book is protected by copyright.
No part of it may be reproduced, stored in a
retrieval system, or transmitted, in any form or by
any means, electronic, mechanical, photocopying,
or recording, or otherwise, without the prior
written permission of the publisher.

Made in the United States of America

Library of Congress Cataloging-in-Publication Data

Gamble, James Gibson.
   The musculoskeletal system.

   Includes bibliographies and index.
   1. Musculoskeletal system. I. Musculoskeletal
system—physiology. WE 102 G191m]
QP301.G35   1987   612'.7   87-16657
ISBN 0-88167-326-9

   The material contained in this volume was submitted as previously
unpublished material, except in the instances in which credit has been
given to the source from which some of the illustrative material was
derived.
   Great care has been taken to maintain the accuracy of the information
contained in the volume. However, neither Raven Press nor the editors
can be held responsible for errors or for any consequences arising from
the use of the information contained herein.

9 8 7 6 5 4 3 2 1

FOR

Jeffrey, Laura, Jamison, Justin, and Terrie

Preface

This book is about basic physiological mechanisms of the musculoskeletal system, and is divided into nine chapters that emphasize major themes: (1) Molecular Biology and Genetics, (2) Subcellular Organelles and Cells, (3) Embryology and Growth, (4) Structural Components, (5) Bone Morphology and Biology, (6) Joints, Synovium, Articular Cartilage, (7) Nerve and Muscle, (8) Calcium, Phosphorus, Metabolic Bone Disease, (9) Inflammation, Immunology, Healing. This book is geared solely for clinicians, not for basic scientists. Most clinicians look at the world of basic science in a special way, perhaps best described by the word practical. They are less interested in some esoteric (although admittedly beautiful) theory than in those concepts and facts that can strengthen the scientific basis of their daily practice. For a clinician, the bottom line is taking care of patients.

Time is a premium quantity for clinicians. Between the demands of patients, family, recreation, etc., they rarely have time for an extended reading of a basic science book. (And if most are like me, they will probably begin to doze off after about 30 minutes anyway.) Accordingly, this book is organized in sections (or "bite-sized chunks") that can be read in a piecemeal fashion for 15 to 20 minutes at a time without the loss of general continuity. The most important boldface terms found in the text are defined in the glossary section at the end of each chapter.

Any book on basic science for clinicians should address the problem brought up by one of our residents: When asked during a case if he knew the zones of articular cartilage, he replied, "I have too many important things to learn." Well in fact, that is right! There are a lot of important things to learn, and with the exponential growth of clinical and basic science, there are more and more of them. Therefore, it is the responsibility of educators to point out the "really important" things and present them in an interesting way. This book is an attempt to do that.

James Gibson Gamble

Acknowledgments

I am indebted to many people who helped make this book a reality. Eugene E. Bleck encouraged me from the start, and as chief of our division, he provided the creative milieu and the most effective possible direction, i.e., leadership by example. Lawrence A. Rinsky, my clinical partner in crime, has helped more than he can realize with suggestions and criticisms; most importantly, he picked up the slack when necessary and never once complained. Vicki Kalen read every page and helped improve the text. Anne Burke did all the art, turning fuzzy ideas into clear illustrations.

I particularly want to acknowledge my secretary, Betty Reid, who not only worked diligently on this manuscript but also managed somehow to do all the other things such as scheduling, research papers, grants, telephones, etc. I also am grateful to Susan Schelly and Cary Tanner for their suggestions. Mary Martin Rogers of Raven Press provided continual encouragement. A special thanks goes to Terrie Gordon Gamble, who sacrificed a lot for "the book."

The English poet and essayist Charles Caleb Colton (1780–1832) wrote "The greatest friend of truth is time, her greatest enemy is prejudice, and her constant companion is humility." In an intensely studied field such as the science of the musculoskeletal system, what seems true and relevant today will inevitably seem less so tomorrow, and I am sure that will be the case with much of the content of this book. However, any areas of neglect or error are solely the responsibility of my oversight and prejudices, and I welcome any criticisms or comments from readers.

Contents

CHAPTER 1: MOLECULAR BIOLOGY AND GENETICS

Introduction to Molecular Biology	1
The Structure of DNA	2
DNA Goes to RNA Goes to Protein	4
The Genetic Code	4
Transcription: DNA Goes to RNA	4
Translation: RNA Goes to Protein	5
Regulation of Gene Expression	7
The Anatomy of Chromosomes	7
Genetics of Some Musculoskeletal Diseases	10
Glossary	13
Bibliography	15

CHAPTER 2: SUBCELLULAR ORGANELLES AND CELLS

Subcellular Structure and Biochemistry Are Universal	16
The Typical Cell	16
The Plasma Membrane	18
Subcellular Organization	20

The Nucleus, 20/The Cytosol, 20/The Endoplasmic Reticulum, 21/The Golgi Apparatus, 22/The Mitochondrion, 23/Lysosomes and Peroxisomes, 24/Cytoskeleton, 25

The Undifferentiated Mesenchymal Cell	25
Chondroblasts and Chondrocytes	26
Osteoblasts and Osteocytes	27
The Osteoclast	28
The Fibroblast	29

xii CONTENTS

Glossary 30
Bibliography 31

CHAPTER 3: EMBRYOLOGY AND GROWTH

The Germ Cell Layers: Endoderm, Ectoderm, Mesoderm 33
The Embryonic Period 35
Axial Skeleton Development 36
The Limbs 41
Joints 44
Bone Formation 45
The Fetal Period 47
Functional Anatomy of the Growth Plate 48
   Morphology of the Growth Cartilage, 49/Metaphysis, 51/Circumferential Structure, 52

Further Growth of Bone (Modeling) 53
Glossary 54
Bibliography 55

CHAPTER 4: STRUCTURAL COMPONENTS

The Extracellular Matrix 57
Collagen: The Major Structural Macromolecule 57
   Structure of Collagen, 58/Molecular Heterogeneity, 59/The Interstitial (Fibrillar) Collagens, 60/Basal Lamina and Cytoskeleton Collagen, 60/ Synthesis, 60/Structure and Function of Procollagen, 62/Posttranslational Modifications and Fibrillogenesis, 64/Disorders of Collagen Metabolism, 65

Proteoglycans 67
   Glycosaminoglycans, 68/Proteoglycan Subunits and Aggregates, 69/ Synthesis, Degradation, and the Mucopolysaccharidoses, 72

Elastin, Fibronectin, and Other Proteins 73
Mineral 76
   Composition of Bone Mineral, 76/Osteoid, 76/Mineralization of Bone, 77

Glossary	78
Bibliography	79

CHAPTER 5: BONE MORPHOLOGY AND BIOLOGY

Bone as Material, a Tissue, an Organ	81
Classification of Bone Tissue	82
Bone Surfaces and Membranes	84
Bone Structure	85
Blood Supply	86
Basic Skeletal Processes	87
Remodeling, 88/Bone Growth Factors, 90/Hormones, 92	
Bioelectricity	92
Fracture Healing	93
External Bridging Callus Healing, 93/Cancellous Healing, 95/Primary Bone Healing, 96/Nonunion, 96	
Bone Grafts	96
Myositis Ossificans	97
Glossary	97
Bibliography	99

CHAPTER 6: JOINTS, SYNOVIUM, ARTICULAR CARTILAGE

Joint Morphology	100
The Synovium	101
Synovial Fluid	103
Synovial Fluid Analysis, 104	
Cartilage Varieties	106
Articular Cartilage	107
Aging of Articular Cartilage, 109/Chondromalacia, 109/Osteoarthritis, 110/Rheumatoid Arthritis, 112	
Articular Cartilage Injury	113

| Glossary | 114 |
| Bibliography | 115 |

CHAPTER 7: NERVE AND MUSCLE

The Neuron	116
Membrane Potential	118
Action Potential	119
Synapses	120
Peripheral Nerves	121
Reflex Arc, 122/Proprioceptors, 122/Muscle Tone, 123/Mechanical Nerve Injury, 124/Neuropathies, 124	
Muscle Structure	125
Muscle Ultrastructure	126
Sarcomere Structure and Function, 128	
Excitation–Contraction Coupling	129
Energy Metabolism	130
Muscle Fiber Type	131
Contractile Properties	132
Effects of Training	133
Muscle Atrophy, 134	
Glossary	134
Bibliography	137

CHAPTER 8: CALCIUM, PHOSPHORUS, METABOLIC BONE DISEASE

Biochemistry of Calcium and Phosphorus	138
Physiology of Calcium and Phosphorus	141
Hormonal Regulation	143
Vitamin D, 143/Parathyroid Hormone, 146/Calcitonin, 149	
Metabolic Bone Disease	149
Osteoporosis, 151/Osteomalacia and Rickets, 153/Osteitis Fibrosa, 154/ Renal Osteodystrophy, 155/Marrow Packing Disorders and Malignancy, 156	

CONTENTS   XV

Abnormal Calcification	156
Glossary	158
Bibliography	160

CHAPTER 9: INFLAMMATION, IMMUNOLOGY, HEALING

Inflammation 161
Mediators of Inflammation, 161/Antiinflammatory Medications, 163/ Chronic Inflammation, 164

Immunology 164
Humoral Immunity, 165/Cell-Mediated Immunity, 167/Major Histocompatibility Antigens, 168

Crystal-Induced Inflammatory Arthritis 169
Gout, 170/Calcium Pyrophosphate Dihydrate Disease, 172/Basic Calcium Phosphate Disease, 172

Healing 173
Wound Healing by First Intention, 173/Wound Healing by Second Intention, 174/Tendon and Ligament Healing, 175

Glossary 176

Bibliography 178

SUBJECT INDEX/179

The Musculoskeletal System
Physiological Basics

1
Molecular Biology and Genetics

INTRODUCTION TO MOLECULAR BIOLOGY

Our ability to understand the genetic code is one of the great scientific achievements of the 20th century. Based on this knowledge, genetic engineering, the isolation and manipulation of genes, is now a reality and undoubtedly will greatly influence the practice of medicine. Already, gene cloning has produced human interferon and growth hormone successfully. Other hormones and bioproducts will soon follow. To appreciate and understand these advances, we must also understand the structure and organization of the gene.

Cell structure and function depend on **deoxyribonucleic acid (DNA)**, the molecule of heredity. This chapter reviews the biochemistry of DNA and shows how the information in DNA is transcribed and translated into structural and enzymatic cellular proteins. We will also discuss the anatomy of chromosomes and the genetics of certain musculoskeletal diseases.

By the end of the 19th century, biologists knew that heredity somehow involved chromosomes, and they postulated that cellular information was inherited in discrete units called genes. Laws of gene inheritance were discovered using the techniques of classical genetics (cross breeding, back crosses, etc.). In 1944, Avery and co-workers showed that genes were actually DNA. In 1953, Watson and Crick worked out the three-dimensional structure of DNA, showing that the polynucleotide double helix (Fig. 1.1) was the molecule of inheritance.

Isolation and purification of bacterial restriction enzymes permitted gene cutting and splicing, and in 1976, new quantitative techniques enabled biologists to identify rapidly and synthesize the chemical sequence of any piece of DNA. With this technology, gene cloning became a reality. Today, a specific gene can be isolated, chemically sequenced, synthesized, and integrated into an organism such as *E. coli*. The bacteria reads the gene as its own and

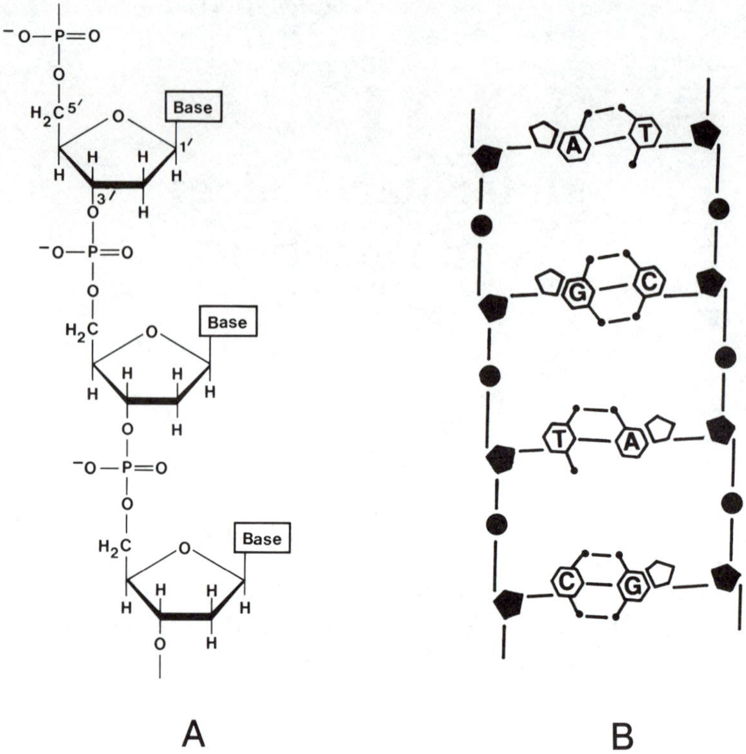

FIG. 1.1. The structure of DNA. **A:** The sugar phosphate backbone is a polymer of 2-deoxyribose to which the four bases—adenine, guanine, cytosine, and thymine—attach at the 1 position. **B:** A skeletal model of the double helix. The four bases are complementary: adenine and guanine on one chain hydrogen bond to thymine and cytosine on the other chain. The turn of the helix repeats every ten nucleotides.

makes the gene product, which can then be purified from the culture medium, ready for use.

THE STRUCTURE OF DNA

The nucleotide sequence of DNA contains all the information needed to assemble and maintain each and every cell in our body. A selective and timely expression of certain genetic programs in the double helix generates the diversity of cell types and extracellular materials.

DNA consists of two long complementary chains of deoxyribonucleotides held together by hydrogen bonds between purine and pyrimidine base pairs. These chains are aggregates of the four deoxyribonucleotides: adenine (A), guanine (G), cytosine (C), and thymine (T). Adenine hydrogen bonds to thymine, and guanine to cytosine, forming complementary polynucleotide chains (Fig. 1.2). Thymine is replaced by the pyrimidine uracil (U) in ribonucleic acid (RNA). A typical cell contains about 6×10^9 deoxyribonucleotides arranged as a double helix in the 46 chromosomes. To put this figure in perspective, one-half of this number, or 3×10^9 letters put into a typical paperback novel would occupy over a million pages.

DNA is packed as a nucleoprotein complex in the cellular nucleus. The exact linear sequence of nucleotides in DNA is the information of the genes. Getting the information out of the DNA involves the use of messenger RNA, transfer RNA, ribosomes, and hundreds of specialized enzymes.

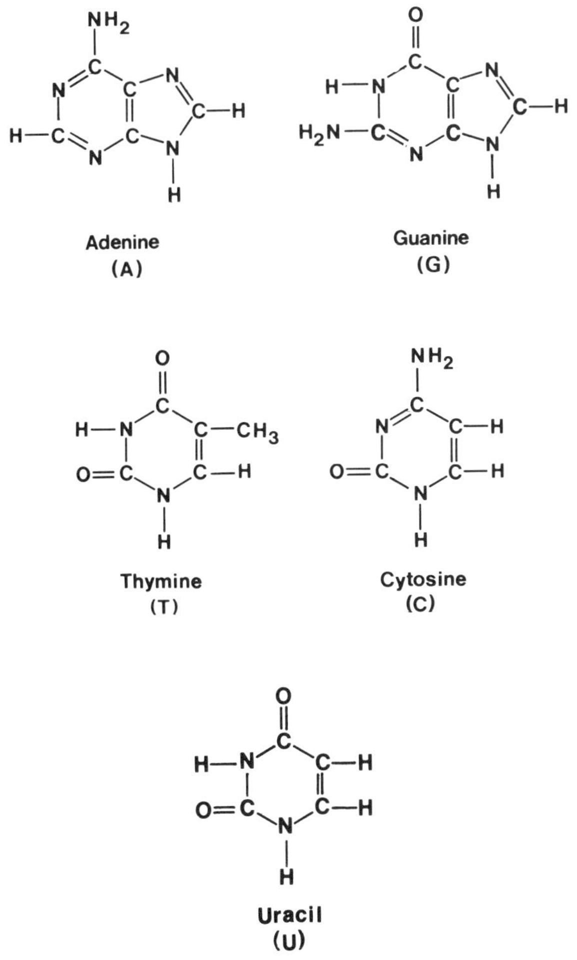

FIG. 1.2. Structure of the organic bases found in nucleic acids. Thymine is replaced by uracil in RNA.

DNA GOES TO RNA GOES TO PROTEIN

The DNA nucleotide sequence of a gene ultimately determines the amino acid sequence of a protein. Genes act by serving as templates for the synthesis of **messenger RNA (mRNA),** and the mRNA is the instruction program for protein synthesis. The process of converting the information in DNA into mRNA is called **transcription.** Control of transcription is by regulatory proteins (repressors) that block or repress DNA until they are removed. After derepression of the gene, the enzyme **RNA polymerase** copies DNA into messenger RNA.

Besides messenger RNA, cells also contain ribosomal RNA and transfer RNA. **Ribosomal RNA** assumes a convoluted shape and combines with 70 different proteins to make up a ribosome, which is the workbench of protein synthesis. **Transfer RNA (tRNA)** carries amino acids to the mRNA–ribosome complex for incorporation into a polypeptide chain. This process is called **translation.** Utilizing transcription and translation, a typical cell synthesizes over 10,000 different proteins.

THE GENETIC CODE

By the end of 1966, Nirenberg, Ochoa, and Khorana had worked out the genetic code of DNA. They discovered that the DNA bases are read as **triplets** (also called **codons**) by the enzyme RNA polymerase. There are 64 possible triplets of the four nucleotides, and since cells use only 20 different amino acids, most amino acids have more than one codon. Actually, 61 codons designate amino acids, and three codons act as termination signals, telling RNA polymerase where to stop. For example, the triplet nucleotide CCC codes for proline, whereas UAG does not code for an amino acid but instead terminates synthesis (Table 1.1).

Exchanging one nucleotide for another in the DNA or shuffling the sequence of codons alters the genetic code. This is the basis of most **mutations,** and even minimal changes have serious consequences. For example, a single codon change in the hemoglobin gene on chromosome 11 causes valine to substitute for glutamic acid in the sixth position of the β chain. This single amino acid switch produces sickle cell anemia.

TRANSCRIPTION: DNA GOES TO RNA

DNA is packaged with specialized regulatory proteins called histones. Removal of histones and other protein repressors permits the enzyme RNA polymerase to copy the DNA sequence into messenger RNA. Recent experiments have discovered a curious fact about the process of transcription. Messenger RNA is not the one-to-one transcription product of DNA; the actual process of transcription is more complicated.

TABLE 1.1. *The genetic code*[a]

First position (5' end)	Second position				Third position (3' end)
	U	C	A	G	
U	Phe	Ser	Tyr	Cys	U
	Phe	Ser	Tyr	Cys	C
	Leu	Ser	Stop	Stop	A
	Leu	Ser	Stop	Trp	G
C	Leu	Pro	His	Arg	U
	Leu	Pro	His	Arg	C
	Leu	Pro	Gln	Arg	A
	Leu	Pro	Gln	Arg	G
A	Ile	Thr	Asn	Ser	U
	Ile	Thr	Asn	Ser	C
	Ile	Thr	Lys	Arg	A
	Met	Thr	Lys	Arg	G
G	Val	Ala	Asp	Gly	U
	Val	Ala	Asp	Gly	C
	Val	Ala	Glu	Gly	A
	Val	Ala	Glu	Gly	G

[a]Sixty-one triplets code for amino acids, and three code for termination signals. The first and second positions of a triplet are relatively invariant, but the third position is less important. In fact, GU plus any one of the four bases codes for valine.

Genes contain intervening sequences of nucleotides that do not show up in the final mRNA. In other words, the actual gene is split by intervening segments of DNA. The complete original RNA transcription product is called **heteronuclear RNA (hnRNA),** and it has to be processed to remove those intervening nucleotide sequences. Walter Gilbert coined the terms describing this situation. The retained segments of a gene are called **exons** because they are *ex*pressed, and the interruptions that split the gene are called **introns** because they are *intra*gene segments. Before RNA leaves the nucleus, the introns are removed, and the exons are spliced together to complete the messenger RNA. This is called RNA processing (Fig. 1.3).

Not all genes are made equal. For example, genes coding for histones and interferon have no introns. On the other hand, the gene for collagen contains at least 50 introns, the most of any gene sequenced so far. The exact role of introns remains speculative. Probably, they help regulate the complex and coordinated nuclear events necessary for transcription and processing.

TRANSLATION: RNA GOES TO PROTEIN

Proteins are the tools of cellular metabolism. They amount to over one-half of all cellular solids, and considerable cellular energy goes into making

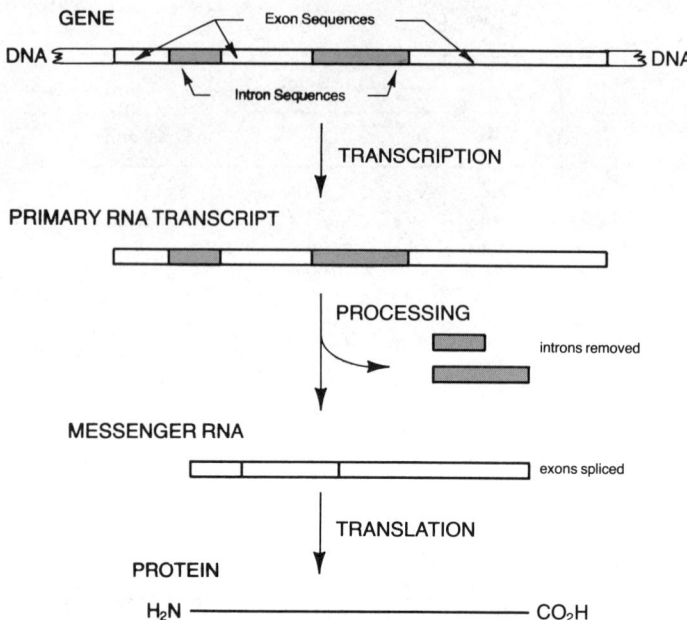

FIG. 1.3. Introns and exons. Most genes are split by intragene sequences (introns), which separate the expressed sequences (exons). The primary RNA transcript (also known as hnRNA) contains both introns and exons. During RNA processing, the introns are removed, and the exons are spliced together, making the messenger RNA. Translation of the mRNA produces the protein coded for by the gene.

them. The fundamental chemical reaction of protein synthesis is peptide bond formation. This reaction requires the interaction of mRNA, tRNAs, and ribosomes.

The central shuttles are tRNAs, to which the various amino acids become attached and thermodynamically activated. Transfer RNA is 75 nucleotides long and has a unique three-dimensional folded structure with a specific terminal attachment site for a specific amino acid. A special set of enzymes called **aminoacyl tRNA synthetases** couple the specific amino acid to the appropriate tRNA, generating an **aminoacyl tRNA.** The aminoacyl tRNA is the activated form of the amino acid, ready to interact with the ribosome and messenger RNA.

A **ribosome** is actually a large multienzyme complex that organizes and catalyzes the events of peptide bond formation. After completion of the mRNA–ribosome complex, the appropriate amino acid is selected by matching up the mRNA codon with the aminoacyl tRNA anticodon. Thus, a ribosome reads the mRNA sequence three nucleotides at a time and selects the appropriate aminoacyl tRNA, making sure the mRNA codon fits the tRNA anticodon.

Three distinct steps actually occur. First, an incoming aminoacyl tRNA binds to the mRNA–ribosome complex by codon–anticodon pairing. Next, a peptide bond forms between the growing polypeptide chain and the incoming aminoacyl tRNA. Finally, the ribosome ejects the empty tRNA and moves three nucleotides down the mRNA molecule to expose a new codon. This process continues until a stop codon turns up; then the polypeptide chain and the ribosome dissociate from the mRNA. In the living cell, these reactions run with incredible efficiency. For example, it takes about 20 seconds to make an average-sized protein of 400 amino acids.

REGULATION OF GENE EXPRESSION

At each step of transcription and translation, feedback controls regulate the rate of these biosynthetic reactions. The most sensitive regulation of protein synthesis involves control of the initiation factors. These are small obligatory proteins without which the mRNA, ribosome, tRNA system will not work. One particular initiation factor called IF2 rapidly inactivates if environmental conditions are unfavorable for protein synthesis. Thus, by a series of finely tuned reactions involving initiating factors, messenger RNA, and aminoacyl tRNAs, each cell manufactures just the right amount of the exact proteins for normal growth, development, and maintenance.

We have seen how the transcription of genetic programs occurs in the cell nucleus and how extensive cleavage and splicing of the hnRNA transcript generates the mRNA program that enters the cytoplasm for translation on ribosomes into protein (Fig. 1.4). Next, we will consider the structure of chromosomes and some genetic diseases.

THE ANATOMY OF CHROMOSOMES

Each human cell contains 46 chromosomes consisting of 22 homologous pairs of autosomes and one pair of sex chromosomes, XX in the female and XY in the male. In the **diploid** cell, one set of chromosomes comes from each parent (the haploid gamete). Chromosomes stained with the Giemsa reagent show distinct bands that permit identification of each chromosome. After staining, the chromosomes can be photographed and paired to show the individual karyotype.

Every chromosome has two **chromatids,** which have comparable genes (Fig. 1.5). These pairs of genes, known as alleles, are not necessarily equal partners. For instance, a single gene mutation may cause an abnormal characteristic despite the presence of a normal partner allele. In this case, the mutant is dominant over the normal allele. If the normal phenotype prevails, the mutation is recessive, and the individual is a carrier.

A **centromere** joins the two chromatids of a complete chromosome. Spin-

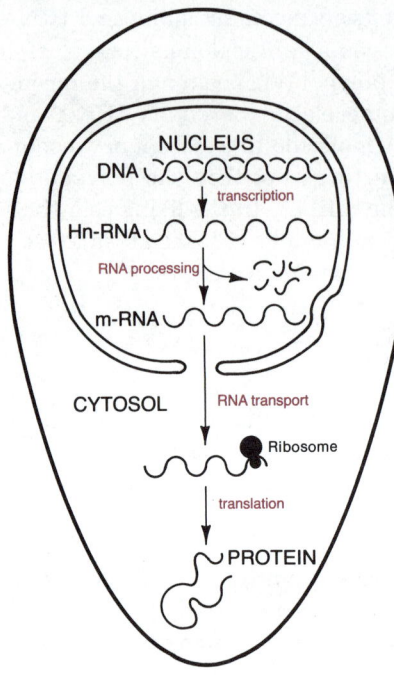

FIG. 1.4. A schematic diagram of transcription, RNA processing, transport, and translation. The nuclear membrane confines transcription to the nucleus and translation to the cytosol.

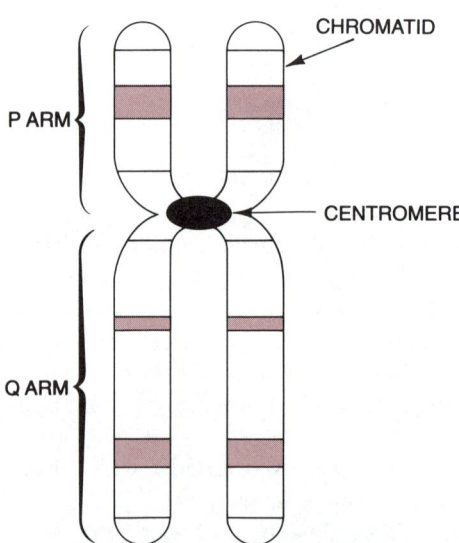

FIG. 1.5. Structure of a typical chromosome. The centromere joins the two chromatids and separates the short p arms from the longer q arms. Certain dyes such as Giemsa reagent produce characteristic and reproducible bands on the chromosomes, which facilitate identification.

dle fibers insert into the centromeres and pull the chromosomes to the cellular poles prior to division. Both gene number and gene balance are important for normal cellular function.

The DNA in a chromosome is a single uninterrupted strand. Since haploid human cells have 3×10^9 nucleotides in 23 chromosomes, each chromosome has 1.3×10^8 nucleotides or close to 5 centimeters of DNA. Obviously, the DNA must be condensed and packed to fit inside the cell nucleus. Actually, several orders of packing are known, and this is accomplished with the help of histones.

The DNA–histone complex is called **chromatin** and is about half histone and half DNA. There are five types of **histones:** H1, H2A, H2B, H3, and H4. They are positively charged proteins that electrostatically bond to the negatively charged DNA (the phosphate groups on DNA give the negative charge).

The first order of packing starts by winding 140 base pairs of DNA around a core of eight histones (two each of H2A, H2B, H3, and H4), making a bead-like particle called a **nucleosome.** Dispersed chromatin is actually a flexible chain of nucleosomes.

At the next level of DNA compaction, the chain of nucleosomes twists into loops of varying size called **looped domains.** The looped domains probably represent functional units of the chromosome: i.e., the whole loop is either activated or repressed. H1 histone and other nonhistone proteins help to pack and maintain this higher order of condensation.

To summarize, DNA is wrapped into nucleosomes, which are packed into a series of looped domains. These domains are further compressed into a compact nucleoprotein comprising the chromosomes (Fig. 1.6).

Chromosomes are unpacked in interphase cells, and DNA is available for transcription. Strangely, not all the genes for a given metabolic pathway are on the same chromosome, so small areas on many separate chromosomes must be unpacked to derepress all the genes of a metabolic sequence. For example, the eight genes for the Krebs tricarboxylic acid cycle enzymes are located on seven different chromosomes. Also, some genes occur in tandem, with multiple gene copies on the same chromosome. This is true for histone genes, tRNA genes, and ribosomal RNA genes. Up to 100 copies of the ribosomal RNA genes cluster on not one but on many different chromosomes. In the nondividing interphase cells, all ribosomal RNA genes from the different chromosomes cluster together in an organized nuclear clump called the **nucleolus.**

In any given cell, less than 10 percent of the chromatin is activated, which means that only 10,000 of the more than 100,000 genes are derepressed. Depending on which genes are turned on, either a chondrocyte or a neuron or any other specialized cell type emerges from the undifferentiated embryonic cell.

Prior to cell division, DNA must be replicated. In eukaryotic cells, repli-

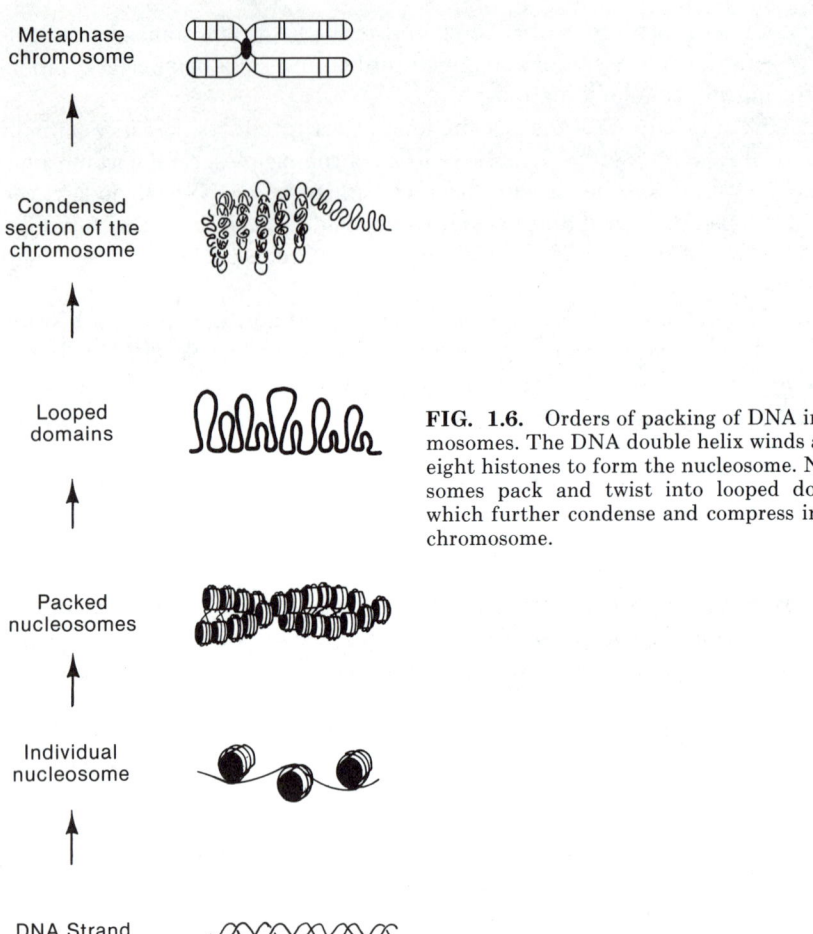

FIG. 1.6. Orders of packing of DNA in chromosomes. The DNA double helix winds around eight histones to form the nucleosome. Nucleosomes pack and twist into looped domains, which further condense and compress into the chromosome.

Prior to cell division, DNA must be replicated. In eukaryotic cells, replication occurs at many sites simultaneously, and all the fragments are eventually bonded together. Over 6,000 replication points per chromosome have been observed. Segregation of these replicated chromosomes into gametes and recombination in the zygote form the molecular basis of classical genetics.

GENETICS OF SOME MUSCULOSKELETAL DISEASES

It is important to appreciate the difference between genetic and congenital conditions. Not all genetic diseases are congenital, and not all congenital syndromes are genetic. For instance, Dupuytren's contracture is genetic (autosomal dominant), but it appears late in life; thus, it is not congenital. Con-

TABLE 1.2. *The three types of genetic diseases and specific examples*

Type of genetic disease	Example
I. Mendelian disorders	
Autosomal dominant	Achondroplasia
Autosomal recessive	Sickle cell anemia
X-linked dominant	Vitamin-D-resistant rickets
X-linked recessive	Hemophilia
II. Chromosomal disorders	
Maldistributions	Down syndrome, trisomy 21
Deletions, translocations	Prader–Willi syndrome
III. Multifactorial disorders	Clubfoot, scoliosis

genital means a deformity present at birth. Phocomelia (hands or feet attached to the body) is a congenital disorder but not genetic, as drugs like thalidomide have demonstrated. Some conditions are both genetic and congenital such as clubfoot or syndactyly. This section presents the three major types of genetic diseases: mendelian disorders, chromosomal disorders, and multifactorial disorders (Table 1.2).

There are four important inheritance patterns of **mendelian disorders:** (1) autosomal dominant, (2) autosomal recessive, (3) X-linked dominant, and (4) X-linked recessive. Autosomal dominant disorders are the most frequent.

Autosomal dominant mendelian disorders occur when only one abnormal gene on one member of the chromosome pair results in the individual being affected. Usually the abnormal gene produces a disease that is either a structural problem or a degenerative condition. Autosomal dominant disease can present as a new gene mutation in a family or can be inherited from an affected parent. Statistically, 50 percent of the children from an autosomal dominant parent are affected, and 50 percent are normal. Either sex may be involved. Table 1.3 lists the autosomal dominant diseases of musculoskeletal importance.

People who have these diseases do not necessarily have identical manifes-

TABLE 1.3. *Autosomal dominant mendelian disorders*

Achondroplasia	Marfan syndrome
Apert syndrome	Multiple exostoses
Camptodactyly	Nail–patella syndrome
Carpal coalitions	Neurofibromatosis
Dupuytren contracture	Osteogenesis imperfecta
Facioscapulohumeral dystrophy	Polydactyly
Gout	Radioulnar synostosis
Malignant hyperthermia	Tarsal coalitions

tations. This is because of the presence of variable expressivity and incomplete penetrance. Variable expressivity occurs when other genetic and environmental influences decrease the expression of a gene's function. Incomplete penetrance results when a dominant gene is present, but the gene is not completely turned on.

Autosomal recessive disorders appear when the abnormal gene must be present on both members of the chromosome pair for the individual to be affected. People who are heterozygotes have only one abnormal gene and are phenotypically normal; they are carriers of the gene. When both parents are carriers, the probability that a given child will have the disease is 25 percent. The probability that a child will be normal is 25 percent, and the probability is 50 percent for a carrier. Autosomal recessive disorders often produce biochemical abnormalities from the absence of a particular enzyme. Table 1.4 lists examples of autosomal recessive disorders.

When an abnormal gene is either on one of the mother's X chromosomes or on the father's single X chromosome, the disease is **X-linked.** The gene can be either dominant or recessive. In **X-linked dominant** conditions, mothers with the gene will have abnormal children 50 percent of the time, and fathers with the gene will have all normal sons and all abnormal daughters. There is no father-to-son transmission of X-linked disease. These diseases include vitamin-D-resistant rickets, pseudohypoparathyroidism, and hypophosphatemia.

In **X-linked recessive** diseases, the mothers are the carriers; 50 percent of the daughters will be carriers, and 50 percent of the sons will have the disease. Affected males have normal sons and carrier daughters. Examples include hemophilia A and Duchenne's muscular dystrophy. Approximately, twenty-five percent of the cases of hemophilia born each year are new mutations, but once established, the condition is propagated in an X-linked recessive pattern.

Chromosomal disorders involve a physical change in the structure of the chromosome. Chromosomal translocations, deletions, and duplications produce recognizable clinical syndromes by shuffling the genetic program or changing the gene balance. Chromosomal maldistributions are more common than previously appreciated. At least 4 percent of pregnancies have chromosomal abnormalities, but most undergo spontaneous abortion; one in 200 live births have chromosomal anomalies.

TABLE 1.4. *Autosomal recessive mendelian disorders*

Chondroectodermal dysplasia	Hypophosphatasia
Diastrophic dwarfism	Metaphyseal dysplasia
Fanconi syndrome	Morquio syndrome
Hurler syndrome	Sickle cell anemia
Gaucher disease	Thalassemia

Extra chromosomal material, as occurs in trisomy, results from nonsegregation or nondisjunction of chromosomes in meiosis. Those trisomes of orthopedic concern include trisomy 18, Down syndrome (trisomy 21), and Klinefelter syndrome (XXY). Trisomy 18 presents with vertical talus, overlapping of the fingers, low-set ears, and micrognathia. Children with Down syndrome have hypotonia and ligamentous laxity with hand, feet, and hip abnormalities. Those with Klinefelter syndrome have elbow and hand dysplasia as well as the characteristic long limbs and hypogonadism.

The most perplexing genetic conditions are the **multifactorial disorders.** Many common musculoskeletal problems are multifactorial. In these conditions, many genes at different loci presumably interact with environmental factors (chemicals, temperature, etc.) to cause the disorder. They do not obey mendelian inheritance rules, and the chromosomal studies are normal. Examples include neural tube defects (myelomeningocele), congenital hip dysplasia, clubfoot, and idiopathic scoliosis.

In these conditions, the risk for first-degree relatives is estimated to be the square root of the incidence of the disease, but the risk increases as other family members are affected. Sex influences the frequency of the disease. For instance, 80 percent of those with severe idiopathic scoliosis are females, whereas rigid clubfoot is twice as frequent in males as in females. The gene threshold for phenotypic expression of many multifactorial diseases is lowered or raised by the presence of the X or Y sex chromosome.

Congenital malformations are gross structural defects apparent at birth, affecting 4 to 6 percent of all liveborn infants. As mentioned previously, not all malformations are genetic or chromosomal. For instance, proximal focal femoral deficiency and congenital scoliosis are nongenetic congenital malformations. Environmental factors such as infectious agents, nutritional deficiencies, or drugs can influence embryogenesis and development, producing congenital malformations. These are discussed more thoroughly in Chapter 3.

GLOSSARY

Allele Alternative forms of a gene at the same gene locus. Homozygotes have the same alleles at the locus of concern, and heterozygotes have two different alleles.

Autosome Any chromosome other than the sex chromosome.

Base pairs The organic bases of DNA that match up by hydrogen bonding: adenine–thymine, guanine–cytosine. In RNA, uracil takes the place of thymine.

Centromere The round, constricted part of the chromosome containing highly redundant DNA where the chromatids are joined and to which the spindle fibers attach during cell division.

Chromatid One of the two elements of the chromosome joined at the centromere divided into a long and short arm in most chromosomes. Chromatids separate during cell division, going to opposite poles of the dividing cell, and each becomes a chromosome of one of the two daughter cells.

Chromatin The DNA–histone complex found in the interphase cell nucleus.

Chromosome A linear thread of DNA that is packed with histones and other proteins, transmitting genetic information and composed of two chromatids.

Clone Identical genes or progeny derived by asexual reproduction from a single gene or single individual.

Codon A group of three bases coding for one amino acid.

Diploid Cells having two sets of chromosomes, as in most cells except gametes and RBCs. The diploid human cell contains 46 chromosomes: 22 homologous autosomes and one pair of sex chromosomes.

Exon That part of the DNA sequence retained in the messenger RNA that codes for the amino acid sequences in a protein.

Gene That part of the DNA that contains the codes for production of a protein or a polypeptide chain. A biologic unit of heredity that is located at a definite position on a particular chromosome.

Genome The total genes of a cell.

Genotype The actual genetic make-up of an individual.

Haploid Containing one set of chromosomes, as in the human gametes, which have 23 chromosomes.

Heteronuclear RNA The immediate one-to-one RNA copy of DNA that is processed by intron excision to make messenger RNA.

Histones The basic (positively charged) proteins that bind to DNA for regulation and condensation.

Intron Also known as intervening sequences, that part of the DNA sequence not present in messenger RNA and thus not expressed in the amino acid sequence of proteins.

Locus The part of a chromosome where a gene is located.

Looped domains A higher order of DNA packing in chromosomes in which the nucleosomes are condensed and packed into loops.

Messenger RNA RNA made by splicing all the exons of a gene; it is the message directing the synthesis of proteins.

Nondisjunction Failure of chromosomes to separate during meiosis giving a daughter cell an extra chromosome.

Nucleosome The unit of first-order DNA packing that is made up of 140 base pairs of DNA wrapped around an octamer of histones.

Phenotype The complete physical and biochemical makeup of an individual, determined by both genetics and environment.

Recombinant DNA A segment of DNA enzymatically inserted into (recombined with) the DNA of another organism that is capable of replication.

Restriction enzymes Bacterial enzymes that cut DNA at specific sites and are used to isolate gene fragments for sequence determination and recombinant techniques.

Reverse transcriptase A viral enzyme that catalyzes the production of a DNA copy of RNA (the opposite enzyme to RNA polymerase).

Ribosomal RNA The 18 S and 27 S RNA that combines with 70 ribosomal proteins to make a ribosome.

Transcription The production of an RNA copy of the DNA.

Transfer RNA The cytoplasmic RNA that specifically binds to and activates amino acids for protein synthesis.

Translation The production of a protein from the nucleotide message in messenger RNA.

Translocations Chromosomal abnormalities in which a part of one chromosome is exchanged for or simply added onto another nonhomologous chromosome.

BIBLIOGRAPHY

Alberts, C., Bray, D., Lewis, J., Raff, M., Roberts, K., and Watson, J. D. (1983): *Molecular Biology of the Cell.* Garland Publishing, New York, London.

Cowell, H. R. (1980): Genetic aspects of clubfoot. *J. Bone Joint Surg.,* 62A:1381–1384.

McKusick, V. A. (1981): The anatomy of the human genome. *Hosp. Pract.,* 16:82–100.

Pembrey, M. E., and Oley, C. (1986): Genetics old and new. *Practitioner,* 230:693–700.

Roberts, F., and Pembrey, M. E. (1985): *An Introduction to Medical Genetics.* Oxford University Press, Oxford.

Smith, D. W. (1982): *Recognizable Patterns of Human Malformation. Genetic, Embryologic and Clinical Aspects.* W. B. Saunders, Philadelphia.

Stryer, L. (1981): *Biochemistry,* second edition. W. H. Freeman, San Francisco.

Wilson, M. G. (1984): Genetics of common orthopaedic congenital malformations. *Contemp. Orthop.,* 8:61–67.

Wynne-Davis, R., Littlejohn, A., and Gormley, J. (1982): Aetiology and interrelationship of some common skeletal deformities. *J. Med. Genet.,* 19:321–328.

2
Subcellular Organelles and Cells

SUBCELLULAR STRUCTURE AND BIOCHEMISTRY ARE UNIVERSAL

Each person has over 200 specialized cell types, different both morphologically and biochemically, to perform the diversity of tasks required in daily living. As divergent as the neuron and the chondrocyte may seem at first, they share a common subcellular structure and biochemistry that are universal among eukaryocytes (nucleated cells). It is the proportion of various subcellular organelles and the utilization of different structural proteins and enzymes that account for cellular diversity. This chapter reviews general subcellular anatomy and demonstrates how certain cells use subcellular organelles and special biochemistry for specific musculoskeletal tasks.

THE TYPICAL CELL

The typical cell has a **plasma membrane** as a semipermeable barrier between the extracellular environment and the intracellular cytoplasm. The cytoplasm contains subcellular organelles, which are membrane-bound compartments responsible for division of labor in the cell. Subcellular organelles of the cytoplasm include (1) nucleus, (2) cytosol, (3) endoplasmic reticulum, (4) Golgi apparatus, (5) mitochondria, (6) lysosomes and peroxisomes, and (7) cytoskeleton (Fig. 2.1).

The largest subcellular organelle is the **nucleus,** bound by a fenestrated nuclear membrane. The nucleus is the genetic control center of the cell and contains chromatin plus the nucleolus. The nucleus ultimately directs cell activity with mRNA.

The **cytosol** is the amorphous liquid in which organelles are suspended. This is the metabolic center, containing most intermediary metabolism

SUBCELLULAR ORGANELLES AND CELLS

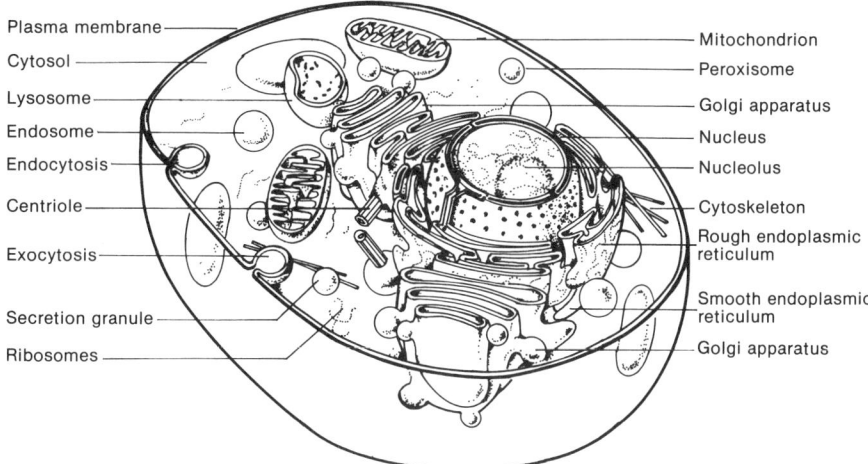

FIG. 2.1. Cytoarchitecture of a typical cell. Adapted with permission from de Duve (1984).

enzymes and also containing the **cytoskeleton,** a collection of radiating filaments that originate near the centriole and project throughout the cell. Along with actin filaments, the cytoskeleton gives shape and motility to the cell and helps transport and rearrange organelles.

The **endoplasmic reticulum** is a network of flat membrane sheets. Rough endoplasmic reticulum, studded with ribosomes, is the site of protein synthesis. Smooth endoplasmic reticulum lacks ribosomes and specializes in synthesis and transport of lipids and steroids.

The **Golgi apparatus** is a convoluted system of membrane sacs that modify, package, and distribute molecules for external secretion and for internal consumption. The **mitochondrion** is the powerhouse of the cell, where substrate oxidation is used to make ATP, the energy currency of the cell. Finally, **lysosomes** contain hydrolytic enzymes and function in intracellular digestion, whereas **perioxisomes** contain specialized oxidative metabolic pathways to rid the cell of toxic peroxides and other chemicals.

Cellular biochemistry revolves around four major biochemical groups: carbohydrates (sugars), lipids (fats), amino acids, and nucleotides. Nucleotides make up DNA and RNA and participate in energy metabolism as ATP and ADP. Amino acids are the subunits of proteins. Lipids contribute to membrane construction and are a source of metabolic energy. Carbohydrates supply most of the energy for the cell and also contribute to such structural macromolecules as proteoglycans.

The central metabolic pathway is **glycolysis** in which glucose is oxidized to pyruvic acid with the formation of small amounts of ATP. In mitochondria, the **Krebs citric acid cycle** oxidizes the pyruvic acid to carbon dioxide and water. Electrons obtained from this process shuttle down the electron transport chain, and the free energy released from the shuttle runs **oxidative phosphorylation,** making ATP. Cells can use ATP for molecular biosynthesis or to energize specialized tasks such as electrical conduction or mechanical motion.

THE PLASMA MEMBRANE

Each histological class of human cells has a unique plasma membrane separating the cytoplasm from the extracellular space. All plasma membranes are sheet-like structures, about 75 angstroms wide, constructed mostly of lipids (phospholipids) and proteins arranged in a mosaic pattern. The lipids assemble as a double layer (a lipid bilayer) with the nonpolar hydrophobic side chains of the lipid toward the inside of the membrane and the polar hydrophilic phosphate groups facing outside. The major membrane lipids are phospholipids, glycolipids, and cholesterol.

Proteins, which vary in size and shape, are dispersed throughout the lipid bilayer (Fig. 2.2). Each variety of cell differs in the amount and kind of proteins. For example, the hepatocyte membrane is 50 percent protein, whereas the Schwann cell myelin membrane is only 18 percent protein. These proteins perform specific functions such as energy transduction, ion transport, or receptor binding. Some proteins span the membrane and undergo allosteric conformational changes when they bind their substrates. These allosteric changes trigger intracellular events that control and alter cellular biochemistry.

Membranes are more a fluid boundary than a rigid structure. Both the lipids and the proteins have considerable lateral mobility in the plane of the membrane. In fact, all membrane molecules move about, having no fixed

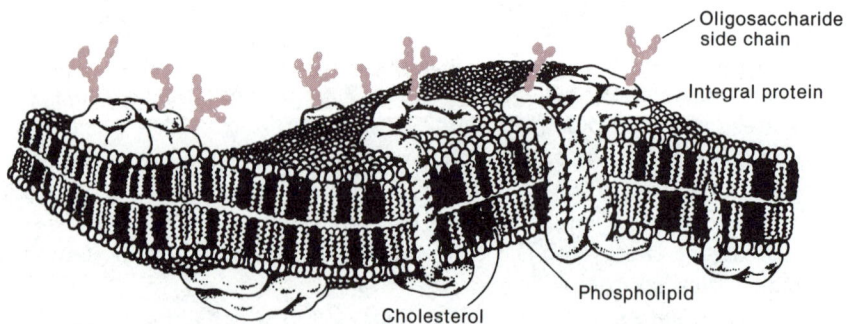

FIG. 2.2. The plasma membrane, a bilayer mosaic of lipids and proteins. Some of the proteins span the bilayer. Oligosaccharides attached to the outside are part of the glycocalyx. Adapted with permission from de Duve (1984).

position in the bilayer except at certain places where the membrane is anchored to the cytoskeleton or to other cells (desmosomes).

Metabolic materials get across membranes in three ways. First, compounds like water, oxygen, and carbon dioxide have **free diffusion** from an area of high concentration to one with a lower concentration. Next, ions like Na^+, K^+, and Ca^{2+} and substrates like glucose and amino acids have a **carrier-mediated active transport,** i.e., transport of the substrate on an allosteric protein carrier that requires energy input from ATP. Finally, large molecules and particulate matter enter cells by the process of **endocytosis.**

Diffusion has been extensively studied, and the governing physical laws, such as chemical weight and fat solubility of the molecule, are well worked out. Generally, the smaller the molecule, the faster the diffusion; the more lipid soluble the molecule, the better the membrane penetration.

Membrane carriers for sugars, amino acids, and other organic substrates have been isolated, characterized, and studied *in vitro*. These carriers are multisubunit protein complexes, some of which form channels, whereas others shuttle back and forth in the bilayer, delivering substrate to the inside.

Some membrane carriers work by exchanging one ion for another. For example, the Na^+-K^+ATPase ejects Na^+ ions from the cytoplasm in exchange for extracellular K^+ ions. This enzyme maintains a high intracellular K^+ concentration and a low Na^+ concentration, establishing a resting membrane potential of about 70 millivolts (negative inside).

Only recently has the mechanism of endocytosis been characterized. Receptor-mediated endocytosis is important in hormone physiology as well as macromolecular ingestion. When a hormone or a macromolecule binds to a membrane receptor protein, a series of membrane-specific events begin, culminating in the invagination of the plasma membrane to form a pit enclosing the receptor–hormone complex. In the case of parathyroid hormone, the binding process immediately triggers the appropriate cellular response of calcium uptake. Then the cell membrane invaginates, putting the hormone and receptor in a vesicle that is digested by the intracellular lysosomal system. In this case, the hormone response is terminated by endocytosis. In osteoclasts, endocytosis brings in other complexes such as partially degraded collagen and mineral residue for a complete lysosomal digestion.

A fluid mosaic lipid–protein bilayer also makes up the membranes of the nucleus, mitochondria, endoplasmic reticulum, Golgi apparatus, and lysosomes. Only the lipid subtypes and protein species change to carry out the special tasks of the organelle.

All cells have a carbohydrate surface coat, the **glycocalyx** (Gr. *kalyx*, husk), attached to the extracellular surface of the bilayer. On electron micrographs, the glycocalyx appears as a fuzzy coat on the outside of the membrane. Some cells have a thick glycocalyx that is recognizable as a basement membrane, whereas other cells like the osteocyte have a glycocalyx that is hardly detectable. The glycocalyx serves to protect cells against certain physical and chemical assaults, and it helps cells to recognize certain neighboring cells and establish connections with them.

SUBCELLULAR ORGANIZATION

The Nucleus

As noted in Chapter 1, the nucleus contains the genetic material and choreographs cellular activities. Molecular messages directing protein synthesis originate and are processed in the nucleus. Cells may have one nucleus (the osteoblast) or many nuclei per cell (the osteoclast); in each case, the nucleus is a highly organized collection of chromatin surrounded by a double membrane or **nuclear envelope.**

The nuclear envelope separates DNA (site of transcription) from the cytosol (site of messenger RNA translation). The outer membrane of the nuclear envelope blends into the rough endoplasmic reticulum, and ribosomes attach to these surfaces. The inner nuclear membrane is covered with an amorphous material called the **nuclear lamina.** This lamina binds to and organizes chromatin. The nuclear envelope has many **nuclear pores,** passageways between nucleus and cytosol. Metabolic traffic across these pores includes steroid hormones, genetic derepressors, nucleotides going in, and messenger RNA, ribosomal subunits, and ADP going out.

The most obvious landmark within the nucleus is the **nucleolus,** site of ribosome synthesis. Here ribosomal RNA genes make ribosomal RNA and assemble the ribosome subunits. Proteins making up the ribosome subunits are imported into the nucleolus from the cytosol and sequentially added to rRNA; the assembled ribosomal subunits then exit the nucleus through the nuclear pores to participate in protein synthesis in the cytosol.

The microscopic appearance of the nucleus gives us general information about cellular activity. For instance, a small, dense nucleus has condensed, inactive chromatin, indicating that the cell is using only a few genes to run a few, select metabolic pathways. However, a large, pale nucleus indicates a synthetically active cell where much of the genetic material is active. Cells containing a large or multiple nucleoli are synthesizing many new ribosomes; a small, condensed nucleolus suggests minimal ribosome content and low rates of protein synthesis. Plasma cells have a small nucleus with a prominent nucleolus, and they have one major function, active synthesis of a specific protein antibody. In contrast, liver cells have a large, pale nucleus, indicating a metabolically active and synthetically diverse cell.

The Cytosol

Intermediary metabolism takes place in the compartment outside the organelles called the cytosol. This represents slightly over half of the total

TABLE 2.1. *The volumes occupied by the intracellular compartments and organelles in the typical cell*[a]

Intracellular compartment	Percentage of cell volume	Number per cell
Cytosol	54	1
Nucleus	6	1
Rough Endoplasmic Reticulum	9	1
Smooth Endoplasmic Reticulum plus the Golgi apparatus	6	1
Mitochondria	22	1,700
Lysosomes	1	300
Peroxisomes	1	400

[a]Adapted with permission from Albers et al. (1983).

cell volume (Table 2.1) and contains thousands of enzymes for hundreds of pathways including glycolysis, gluconeogenesis, and lipid and protein synthetic reactions. Historically, this region was called the cellular "ground substance."

By weight, 70 percent of the cytosol is water, and 20 percent is protein (ions, metabolites, and vitamins make up the remainder). The cytosol is not a solution at all but a highly organized gel that displays viscoelastic properties.

Suspended within the cytosol are fat globules, glycogen, and other storage granules, which provide a reserve source of fuel. Cytoskeleton filaments and their associated actin fibers spread throughout the cytosol as an internal scaffolding, giving shape to the cell and spatial organization to all the subcellular organelles.

Separation of cellular biochemistry into reactions located in distinct organelles and those in the cytosol has a great benefit. This permits many incompatible or competitive biochemical reactions to proceed simultaneously. For instance, protein synthesis takes place at the rough endoplasmic reticulum while protein degradation occurs within the lysosomes. Lipid oxidation proceeds in the mitochondria while lipid synthesis occurs at the smooth endoplasmic reticulum. This biochemical compartmentation greatly increases metabolic diversity and versatility over that possible in a one-compartment organism like *E. coli*.

The Endoplasmic Reticulum

The endoplasmic reticulum is a series of membrane layers, extending throughout the cytoplasm, responsible for most protein and lipid synthesis. These membrane layers form a continuous sheet and enclose an irregular sac or cisternal space. The rough endoplasmic reticulum has ribosomes attached to the outside at specific ribosomal binding sites. Many of these ribosomes are attached in such a way that they pass their newly synthesized proteins directly through the membrane and into the cisternal space. In this way, proteins destined for export or for special cellular distribution become segregated from the cytosol and sent on their way to the Golgi apparatus.

Membrane-bound ribosomes come from a pool of unattached ribosomes in the cytosol. Ribosomes can work on messenger RNA one at a time, or many ribosomes can bind in tandem to the same messenger, in which case it is called a polysome. In either situation, proteins destined to remain in the cytosol are synthesized by unattached ribosomes or polysomes, and those to be secreted or sequestered are synthesized by membrane-bound ribosomes on the rough endoplasmic reticulum.

Smooth endoplasmic reticulum is physically continuous with the rough endoplasmic reticulum, but it lacks ribosomal binding sites. One clue to the role of smooth endoplasmic reticulum came with the observation that lipid-producing cells have an abundance of smooth endoplasmic reticulum. Subsequently, the enzymes for steroid metabolism, lipoprotein biosynthesis, and drug detoxification have been found in the smooth endoplasmic reticulum. Also, most of the lipids for cellular use, including phospholipids, glycolipids, and cholesterol, originate from the smooth endoplasmic reticulum.

The Golgi Apparatus

The Golgi apparatus is a collection of membrane stacks, arranged as a series of regular cavities, situated near the nucleus. It functions to sort, con-

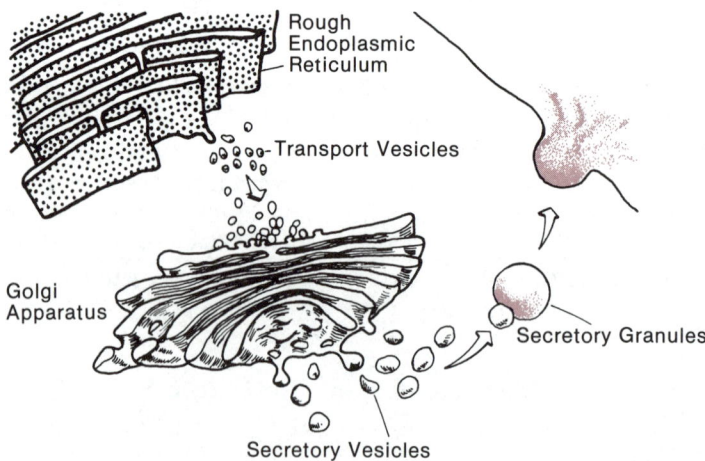

FIG. 2.3. Proteins synthesized at the rough endoplasmic reticulum are transported to the Golgi, where they are processed and packaged for secretion or for internal use. Adapted with permission from Rogers (1983).

centrate, and modify molecules such as enzymes, polypeptide hormones, glycoproteins, and proteoglycans (Fig. 2.3).

Newly synthesized polypeptides and proteins are transferred from the endoplasmic reticulum to the Golgi cisternal cavities in special packets called coated vesicles. Once inside the cisternal cavities, Golgi-specific enzymes carry out chemical modifications such as proteolysis or glycosylation. This final modification is a type of "molecular ticket," permitting the identification and segregation of mature proteins, which are then packed in a secretory vesicle and shuttled to their ultimate intracellular or extracellular destination.

The Mitochondrion

As noted in Table 2.1, mitochondria occupy close to one-quarter of the cytoplasm, and a typical cell contains about 1,700 mitochondria. This large number reflects the importance of mitochondria, which make most of the ATP and help regulate intracellular calcium metabolism.

The mitochondrion can be thought of as a sack within a sack (Fig. 2.4). The outside sack, or **outer mitochondrial membrane,** is a fenestrated barrier that excludes proteins with a molecular weight greater than 10,000 (e.g., the enzymes of the cytosol, macromolecular complexes, and storage granules). The space between the inner and outer membranes is important in the proton flux that is part of oxidative phosphorylation.

The **inner mitochondrial membrane** is a highly convoluted sack with a large surface area, containing ion transport proteins, enzymes of the **respiratory transport chains,** and the **ATP synthetase** system. The matrix space (enclosed by the inner membrane) contains Krebs cycle enzymes,

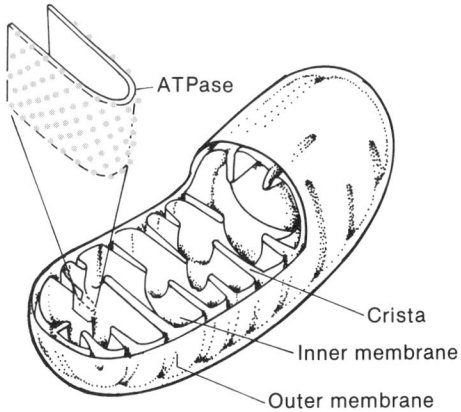

FIG. 2.4. Ultrastructure of the mitochondrion showing ATPase enzyme of oxidative phosphorylation on the inner wall of the inner membrane. Adapted with permission from de Duve (1984).

which metabolize end products of intermediary metabolism such as pyruvate, ketone bodies, and certain amino acids.

Fatty acids and glucose are the major cellular energy sources, and they are ultimately converted to acetyl coenzyme A, which directly enters the Krebs cycle. One spin around the cycle converts acetyl CoA to carbon dioxide, aqueous hydrogen ions, and electrons in the form of NADH and FADH. This is the starting point for ATP production, a process called **chemiosmosis**.

Reduced pyridine nucleotides (NADH, $FADH_2$) feed their electrons into the cytochromes of the electron transport chain. Free energy is released by passage of the electrons down the electropotential gradient of the cytochromes. The inner membrane uses this free energy to pump protons out of the matrix and into the intermembrane space, establishing an electrochemical gradient across the inner membrane (electrons inside and protons outside). The enzyme **ATP synthetase** then uses this electrochemical gradient to make ATP from ADP and Pi. Finally, ATP exits the mitochondria to be used in any energy-requiring reaction.

The complexity of this system is more than justified by the efficiency. Conversion of glucose to pyruvic acid in the cytosol generates a net total of two ATP molecules. The metabolism of pyruvic acid to carbon dioxide and water in the mitochondrion generates 34 molecules of ATP. In general, about 20 percent of ATP production comes from glycolysis in the cytosol, and 80 percent from oxidative phosphorylation in mitochodria. This is certainly a metabolically advantageous organelle, and it is understandable why tissues involved in mechanical work (tissues like the soleus muscle of marathon runners) have an abundance of mitochondria.

Lysosomes and Peroxisomes

The smallest cellular organelles are potentially the most destructive. **Lysosomes** are small membrane-bound particles containing digestive enzymes. The older literature refers to lysosomes as "suicide bags" because it was noted that in programmed cellular death, as occurs in embryogenesis, lysosomes released their enzymes into the cytosol with deadly results. Since then, a much greater role in cellular economy has been unraveled.

Lysosomes digest any material brought into the cell by endocytosis, and they also function to remove worn-out membranes and organelles. This is accomplished with an armamentarium of over 50 degradative enzymes including glycosidases, proteases, nucleases, lipases, phosphatases, and sulfatases. These enzymes are sufficient to degrade all known biological macromolecules into basic monomers such as glucose, glycine, adenine, and palmitic acid. The most favorable condition for these degradative enzymes is an acidic environment near the pH of 5, a condition maintained inside the lysosome. It is speculated that these enzymes have such a low pH optimum to protect the cell against the occasional inadvertent leak of a few of these digestive enzymes into the cytosol, where the pH is close to neutral.

Lysosomes originate from the Golgi apparatus after being loaded with enzymes. They bud off from the Golgi membrane and are called **primary lysosomes.** Recall that macromolecular materials gain entrance into the cell in endocytotic vesicles. It should also be noted that worn-out membranes and organelles are encapsulated in membrane-bound vesicles called autophagic vesicles. Primary lysosomes patrol the cytoplasm until they find endocytotic or autophagic vesicles, whereupon the membranes fuse, and the degradative enzymes encounter the vesicular material and begin the process of digestion. The particle is now called a **secondary lysosome.** Usable products of digestion ultimately diffuse out of the secondary lysosome to enter the mainstream of intermediary metabolism; undigested debris remains inside the now condensed vesicle as a residual body.

Certain inherited diseases result from the absence of a specific lysosomal enzyme. For instance, patients with Hurler disease have bone deformities and organomegaly resulting from the accumulation of intracellular glycosaminoglycans. These people lack iduronidase, an enzyme that is necessary for glycosaminoglycan degradation. Another example is Tay–Sachs disease, in which gangliosides accumulate because of the absence of the enzyme hexosaminidase. A focus of current research in this field involves attempts to provide either the missing gene or the appropriate enzyme to these patients.

Peroxisomes are another group of small membrane-bound particles found in the cytoplasm. They contain enzymes such as D-amino acid oxidase, urate oxidase, and nonspecific oxidases that perform oxidative detoxification reactions. Hydrogen peroxide is a metabolic waste product of these reactions. To eliminate this highly toxic compound, 40 percent of the enzyme content of the peroxisome is catalase, an enzyme that degrades hydrogen peroxide to water.

Cytoskeleton

The shape of a cell is determined by an intracellular scaffolding of crisscrossed protein fibers and tubules known as the cytoskeleton. The fibers vary in thickness from 6 to 20 nm and are rapidly assembled or disassembled, depending on the support needs of a cell. Microtubules attach to mitochondria and other organelles. They have been shown to move the mitochondria to different intracellular locations, presumably closer to sites of energy need. The mitotic spindle is also composed of microtubules and performs the separation of chromosomes during mitosis.

THE UNDIFFERENTIATED MESENCHYMAL CELL

With the exception of osteoclasts, musculoskeletal cells originate from **undifferentiated mesenchymal cells.** As the **progenitor cells,** all mesenchymal cells are derived from the mesoderm. Depending on the timing and the embryological environment, mesenchymal cells differentiate into cartilage, bone, fibrous tissue, muscle, or certain hematopoietic cells.

The structure of the mesenchymal cell reflects its primal function, which is to divide and then to differentiate. It is a shapeless, primative-looking cell. The cytoplasm contains relatively few organelles, the nucleus is pale-staining, and the metabolism is geared for cellular division.

Embryos contain the most undifferentiated mesenchymal cells per unit structure. But even in skeletally mature individuals, these cells remain along the inner surface of the periosteum or close to blood vessels, ready for division and differentiation. At this age they are thin and elongated, resembling endothelial cells or fibroblasts.

CHONDROBLASTS AND CHONDROCYTES

Undifferentiated mesenchymal cells that find themselves in an environment calling for cartilage produce a new population of cells called chondroblasts. **Chondroblasts** are immature chondrocytes not yet surrounded by a cartilage matrix. They are best observed under the perichondrium or in the zone of Ranvier at the growth plate. As the chondroblast matures and secretes extracellular matrix, the cell becomes more differentiated and occupies a pocket in the matrix. That pocket ultimately becomes a lacuna and completely isolates the cell, now a mature chondrocyte, from other cells and from the vascular system.

Chondrocytes make hyaline cartilage, fibrocartilage, elastic cartilage, and epiphyseal cartilge. Precise cellular morphology varies slightly depending on which cartilage is made and where in the tissue the cells are found. For instance, the surface cells of articular cartilage are elongated and flattened, but the intermediate cells are plump and round. The reserve chondrocytes of the growth plate are small and ovoid, but the cells of the hypertrophic zone are large and round. All chondrocytes are isolated cells that lie in lacunae and reside in a hypovascular environment lacking lymphatics and nervous innervation. Chondrocytes get their nourishment by diffusion of substrates through the extracellular matrix. In older people, chondrocytes occasionally divide and form nests of two to six cells in a single lacuna.

Chondrocytes have cell processes that extend a short distance into the matrix but do not touch other cells. This ruffled or scalloped plasma membrane reflects a constant and active cell secretion process necessary to synthesize and maintain the extracellular matrix.

Chondrocytes vary in their synthetic activities; certain cells specialize in the synthesis of type II collagen, whereas others make mostly reticulin fibers or elastin. Ultrastructure varies with the type of material made. For instance, the hyaline chondrocyte has scant rough endoplasmic reticulum but a large Golgi. This cell makes type II collagen and copious amounts of proteoglycans, responsible for metachromatic staining with toluidine blue. On the other hand, the fibrocartilage chondrocyte has abundant rough endoplasmic reticulum. The elastin-secreting chondrocyte has a cytoplasmic storage area for a precursor of elastin (proelastin).

The intracellular events are similar in all chondrocytes after selective activation of appropriate genes for cell-specific proteins. The cells sequester amino acids, which are polymerized into procollagen, proteoglycan cores, and other proteins. Simultaneously, polysaccharide glycosaminoglycans and hyaluronic acid are synthesized. All the molecules are secreted through the plasma membrane secretory vesicles into the extracellular space, where they undergo maturation and assembly. It is the composition of this extracellular matrix and not the ultrastructure of the cell that best characterizes the subpopulation of chondrocytes. The matrix of hyaline cartilge is composed mostly of type II collagen, sulfated glycosaminoglycans, chondroitin-4-sulfate, chondroitin-6-sulfate, and other minor collagen subtypes plus a small amount of elastin.

OSTEOBLASTS AND OSTEOCYTES

Under the appropriate conditions, undifferentiated mesenchymal cells differentiate into **osteoblasts.** These plump polygonal or cuboid cells form a continuous layer on a new-bone-forming surface. The function of the osteoblast is to synthesize and secrete osteoid, the organic matrix of bone. Osteoblasts in lamellar bone lay down successive sheets of osteoid in different orientations, increasing the shear and bending strength of bone. Mineralization begins after extracellular maturation of osteoid, usually 7 to 10 days following deposition.

Osteoblasts are not isolated cells like chondroblasts. The plasma membrane has numerous cytoplasmic processes extending throughout the osteoid, which contact reciprocal cytoplasmic processes from deeper osteoblasts and osteocytes. These processes make permanent connections via tight junctions and assure cellular communication channels deep within the bone. The tunnels through which these processes run are the canaliculi of mature bone.

The ultrastructure of osteoblasts reflects their active metabolic role in osteoid synthesis. The nucleus is large and has one to three nucleoli; the cytoplasm stains intensely basophilic because of the large concentration of ribosomes. In fact, the most obvious subcellular organelles are the extensive rough endoplasmic reticulum and a well-developed Golgi apparatus. Mitochondria are numerous and contain many calcium phosphate granules. Few lysosomes or storage granules are present, and the concentration of the enzyme alkaline phosphatase is quite high. Components of osteoid, such as collagen, proteoglycans, and other noncollagen proteins, are synthesized and rapidly secreted into the extracellular space. They are not saved or stored inside the cell in secretory granules.

An osteoblast becomes an **osteocyte** after it is totally surrounded by mineralized osteoid. Both structure and metabolic activity change as the osteocyte matures in its lacuna. The surface plasma membrane becomes more regular, interrupted occasionally by cellular processes extending into the depths of canaliculi. Osteocytes have much less cytoplasm and fewer organelles than osteoblasts. The nucleus occupies the majority of the cytoplasmic space, and most of the DNA is condensed into inactive heterochromatin. A nucleolus is usually absent, indicating little new ribosome synthesis. The extensive endoplasmic reticulum of the osteoblast is gone, and a low basal metabolic activity maintains local homeostasis.

Osteocyte plasma membranes are not tightly pressed up against the mineral walls of the lacunae. Cells are separated from the walls by a narrow amorphous zone constituting the bone–fluid space. This extravascular space is particularly rich in ions such as potassium and calcium, and it is an important site in mineral homeostasis.

Although the osteocyte may be buried deep in the skeleton, it is not an isolated cell like the chondrocyte. Besides the cellular processes that contact many other cells by tight junctions, the osteocyte receives nutrients, gases, and hormonal messages by diffusion through the extracellular fluid that circulates from the perivascular spaces of the Haversian canal, through the canaliculi, and into the lacunae. In this way, PTH, calcitonin, other hormones, and chemicals influence the osteocyte.

One of the physiological benefits of this system involves short-term calcium regulation. The collective population of osteocytes is large, and the surface area of mineral exposed to the osteocytes is enormous. Under the influence of PTH, the osteocytes can rapidly mobilize surface calcium of the lacunae and quickly deliver it to the circulation. This process is termed **osteocytic osteolysis** and is an important source of rapidly available plasma calcium (see Chapter 8).

THE OSTEOCLAST

Bone resorption is the specialty of **osteoclasts.** These gigantic, motile cells are found where bone is being resorbed during normal remodeling or under a pathological influence. Osteoclasts are multinucleated, some containing several hundred nuclei, and they are found in excavation pits of bone called **Howship's lacunae.**

Unlike other musculoskeletal cells, the osteoclast is not derived directly from the undifferentiated mesenchymal cell. The osteoclast comes from the **lymphoid cell line,** which includes such cells as peripheral blood monocytes and tissue macrophages. From an extraskeletal origin in the spleen and liver, osteoclast precursor cells migrate to the bones via blood vessels. The precursors proliferate, differentiate, and fuse with other osteoclasts to make even larger cells. The exact mechanism and even the reason for this cellular fusion to make a multinucleated, macrocytic cell remain a mystery. What is better understood is the mechanism of bone resorption by osteoclasts.

Bone resorption begins with osteoclast activation by parathyroid hormone or by other proteins such as osteoclastic activation factor. The cytosol of an activated osteoclast contains numerous mitochondria and many secretory granules full of degradative enzymes. Each nucleus is surrounded by a well-developed Golgi apparatus. Most characteristic of active osteoclasts is a **ruffled border** consisting of hundreds of microvilli situated at the resorption surface of the plasma membrane.

Electron micrographs of bone being resorbed show a frayed surface of free collagen ends with bits and pieces of mineral debris. Microvilli of the ruffled border are motile and sweep across the surface of the bone during resorption. Organic acids and packets of enzymes are released from the microvilli, enzymatically digesting the bone. Degradation products reenter the cell by endocytosis at the microvilli surface. Further digestion in secondary lysosomes results in eventual discharge of ions and amino acids back into the general metabolic pool.

Osteoclasts acidify their environment and secrete degradative enzymes to dissolve bone mineral and matrix. The microenvironment of the resorption surface is special. The pH is low, accommodating the degradative enzymes, which function best in an acidic environment. Release of organic acids and carbonic acid into the area guarantees an acidic pH. To keep the digestion confined to the resorption surface, the osteoclast surrounds the area with another membrane modification called the **clear zone.** This is a modification of the plasma membrane that acts as a circumferential suction cup, sealing off the resorption surface from the surrounding space.

In summary, osteoclasts begin in the lymphoid cell line, morphologically resembling monocytes. Proliferation and fusion generate a giant, multinucleated osteoclast. Bone resorption, starting in a Howship's lacuna and progressing to a cutting cone, burrows for a variable distance, and then the osteoclast disappears. Almost immediately on the trail of the osteoclast, osteoblasts follow to begin the process of new bone formation in the cavity left by the osteoclast. Exactly what determines the depth of the cutting cone or in which direction it travels remains a mystery. Bioelectrical phenomena seem to play a role and are considered in Chapter 5.

THE FIBROBLAST

The cell most commonly associated with connective tissue is the fibroblast. It is found in loose and dense connective tissue, in the outer layers of the periosteum, in ligaments and tendons, and anywhere the organism has a need for collagen, for instance, in wound healing.

The fibroblast is a large, elongated, **fusiform** or spindle-shaped cell containing a central ovoid nucleus with finely dispersed chromatin. A well-developed rough endoplasmic reticulum is present, as is a large Golgi complex with numerous secretory vesicles. The cytosol is scant, containing no glycogen granules or fat globules. Numerous branching cytoplasmic processes contribute to the fusiform appearance. The long axis of a fibroblast is aligned with the direction of local collagen fibrils.

The biochemical machinery is geared to collagen and proteoglycan biosynthesis. Scant cytoplasm with extensive rough endoplasmic reticulum and a large Golgi indicate a steady synthesis of proteins, which are continually secreted, not stored. Like chondrocytes, not all fibroblasts are identical. Fibroblasts in tendons and ligaments make type I collagen, whereas perivascular fibroblasts make type III collagen. Selective gene derepression determines both the morphological and the biochemical phenotype of the fibroblasts.

Fibroblasts retain the capacity for growth and regeneration and are involved in physiological as well as pathological healing and in the inflammatory response (Chapter 9).

GLOSSARY

Chemiosmosis The mechanism of oxidative phosphorylation wherein energy of the respiratory transport chain is utilized to make ATP by separating protons and electrons.

Cytoskeleton A collection of fibers and microtubules crisscrossing the cytosol, giving support to the cell and providing a mechanism for moving and distributing organelles within the cytosol.

Cytosol The gel-like interior of a cell containing substrates, enzymes of glycolysis, unbound ribosomes, and the subcellular organelles.

Endocytosis A mechanism by which cells import materials into the cytosol by wrapping them in a sealed vesicle that originates by invagination of the plasma membrane.

Glycocalyx The extracellular carbohydrate coat of a cell, serving for protection and recognition.

Golgi apparatus Site of concentration, modification, and packaging (into secretory granules) of newly synthesized proteins transferred from the rough endoplasmic reticulum.

Lysosomes Membrane-bound bags derived from the Golgi apparatus containing over 60 acidic hydrolytic enzymes, which participate in a wide range of degradative activities. Primary lysosomes are fresh vesicles, recently released from the Golgi. Secondary lysosomes form after primary lysosomes fuse with another vesicle.

Mitochondria The energy producers of the cell, utilizing an orderly sequence of membrane-bound enzymes to generate adenosine triphosphate (ATP) by the process of oxidative phosphorylation.

Nuclear envelope A double membrane system that encloses the nucleus and separates it from the cytoplasm. Within the membranes are nuclear pores, which form conduits for nuclear–cytoplasmic exchange of molecules.

Nucleolus Site for production of rRNA, which is complexed to numerous proteins and then shipped to the cytoplasm for incorporation into ribosomes.

Peroxisomes Specialized membrane-bound small compartments containing the enzyme catalase and other oxidases such as D-amino acid oxidase and urate oxide.

Plasma membrane A selectively permeable fluid mosaic of lipids and proteins separating the cell from the extracellular space.

Respiratory transport chain A series of mitochondrial enzymes, containing cytochromes (c, c_1, b, a, a_3), that transfer electrons produced from substrate oxidation to molecular oxygen, releasing free energy in the process.

Ribosome A ribonucleoprotein complex made up of RNA molecules complexed with specific proteins, which catalyzes mRNA translation.

Rough endoplasmic reticulum A network of ribosome-studded membrane canals where proteins are synthesized.

Ruffled border The resorption part of an osteoclast membrane consisting of microvilli to discharge acids and enzymes and to reabsorb the digestion products.

Smooth endoplasmic reticulum Branching network of membranous tubules containing processing enzymes (e.g., for steroid hormone synthesis) and serving other specialized roles (e.g., uptake and release of Ca^{2+} ions in striated muscle cells).

BIBLIOGRAPHY

Albers, B., Bray, D.,Lewis, J,. Raff, M., Roberts, K., and Watson, J. (1983): *Molecular Biology of the Cell.* Garland Publishing Inc,. New York.

Dautry-Varsat, A., and Lodish, H. F. (1984): How receptors bring proteins and particles into cells, *Sci. Am.,* 250:52–58.

de Duve, C. (1984): *A Guided Tour of the Living Cell,* pp. 100–250. Scientific American Books, New York.

Hancox, N. M. (1972): The osteoclast. In: *The Biochemistry and Physiology of Bone,* edited by G. H. Bourne, pp. 45–67. Academic Press, New York.

Hancox, N. M. (1972): *Biology of Bone.* Cambridge University Press, Cambridge.

Jowsey, J. (1977): Bone morphology: Bone cells. In: *Metabolic Diseases of Bone,* edited by J. Jowsey, pp. 58–83. W. B. Saunders, Philadelphia.

Kosher, R. A. (1983): The chondroblast and chondrocyte. In: *Cartilage, Vol. 1: Structure, Function, Biochemistry,* edited by B. K. Hall, pp. 59–86. Academic Press, New York.

Kreil, G. (1981): Transfer of proteins across membranes. *Annu. Rev. Biochem.,* 50:317–325.

Marks, S. C. (1984): Congenital osteopetrotic mutations as probes of the origin, structure, and function of osteoclasts. *Clin. Orthopaed.*, 189:239–263.
Matthews, J. L. (1980): Bone structure and ultrastructure. In: *Fundamental and Clinical Bone Physiology*, edited by M. R. Urist, pp. 21–44. J. B. Lippincott, Philadelphia.
Nomura, M. (1984): The control of ribosome synthesis. *Sci. Am.*, 250:102–114.
Rogers, A. W. (1983): *Cells and Tissues*. Academic Press, New York.
Rothman, J. (1981): The Golgi apparatus: Two organelles in tandem. *Science*, 213:1212–1216.
Sheldon, H., and Jaeger, V. (1982): The mesenchymal cell—its origin, structure, and function. In: *The Musculoskeletal System. Embryology, Biochemistry, and Physiology*, edited by R. L. Cruess, pp. 17–29. Churchill Livingstone, New York.
Unwin, N., and Henderson, R. (1984): The structure of proteins in biological membranes. *Sci. Am.*, 250:78–94.
Vaughan, J. M. (1975): *The Physiology of Bone*. Clarendon Press, Oxford.

3
Embryology and Growth

Combination of the oocyte and the sperm during fertilization reconstitutes the diploid genotype and initiates the events of embryogenesis. Thirty-eight weeks (or 266 days) later, a child is born consisting of billions of specialized cells that trace their origin back to the one-celled zygote.

Traditionally, gestation is divided into the embryonic and the fetal periods. The major events of the embryonic period concern organogenesis, whereas the fetal period is mostly a time of growth and differentiation. Each step of embryogenesis requires exact timing and control as the genetic messages in the chromosomes unfold.

Within 3 weeks, three embryonic germ layers appear, and by 3 months, all the organ systems are present. Anatomists and embryologists have meticulously documented the details of gestation, and from this wealth of information, certain concepts and specific facts have emerged that help explain normal embryology as well as congenital deformities. This chapter reviews some of the more important embryological concepts in order to provide a framework for understanding the facts of musculoskeletal development.

THE GERM CELL LAYERS: ENDODERM, ECTODERM, MESODERM

Thirty hours after fertilization, cell division begins, and by 3 days the **zygote** has grown into a ball of cells called the **morula** (L. mulberry). A cavity appears in the center of the morula, and the resulting hollow ball, now called a **blastocyst,** has two distinct cell types, the trophoblast and the embryoblast. The trophoblast invades the uterus and forms the placenta. The embryoblast forms the inner cell mass that ultimately gives rise to the embryo.

By the eighth day, the inner cell mass has two cellular layers, the **endoderm** (Gr. *endon,* within), which is a layer of small cuboidal cells, and the **ectoderm** (Gr. *ektos,* outside), which is a layer of columnar cells. This bilaminar germ disk grows until the third week, when an irregularity called the **primitive streak** appears on the ectodermal surface. Cells from the primitive streak divide, invaginate, and migrate between the endoderm and ectoderm, forming the third germ layer, the **mesoderm** (Gr. *mesos,* middle).

Concept 1: The mesoderm gives rise to most of the musculoskeletal system.

By the 15th day, the mesoderm cell layer is well established, and cells from *Hensen's node,* located at the most cephalic extension of the primitive streak, begin another invagination and migration to generate the notocord. The notocord has two basic functions. It is the central axis around which the somites and the spine organize from the mesoderm, and it induces the overlying ectoderm to generate the neural groove, precursor of the brain and spinal cord (Fig. 3.1).

Concept 2: The ectoderm gives rise to structures maintaining contact with the outside world: the central nervous system (CNS), peripheral nervous system (PNS), ears, eyes, nose, hair, nails. The endoderm forms the remaining organs and glands.

Under the inductive influence of the notocord, the neural groove invaginates and closes into the neural tube. Around the 24th day, the mesoderm along both sides of the notocord collects into discrete masses called somites. The first somites develop in the cephalic region, and subsequent somites appear in sequence from the cranial to the caudal region. The somites are recognizable as bulges along the dorsal embryonic surface (Fig. 3.2). Eventually 42 or 44 pairs of somites appear (some embryos have an extra pair of coccygeal somites).

Most of the axial skeleton and associated muscles as well as the dermis of the skin are derived from the somites. This is accomplished by separation of the cells of each somite into three zones, the sclerotome, the myotome, and the dermatome. The sclerotome produces fibroblasts, chondroblasts, and osteoblasts. The myotome differentiates into the segmental muscles of the somite. The dermatome forms the segmental dermis and subcutaneous tissue; the overlying epidermis comes from the ectoderm.

Concept 3: Somites (mesodermal derivatives) give rise to the sclerotome, the myotome, and the dermatome, which in turn generate the skeleton, the muscles, and the dermis.

By the end of the fourth week, most of the somites are present, and cellular activity of the primitive streak has diminished. The streak normally undergoes complete regression. If remnants persist, they can give rise to sacrococcygeal teratomas.

Concept 4: The embryo develops in a craniocaudal as well as a proxi-

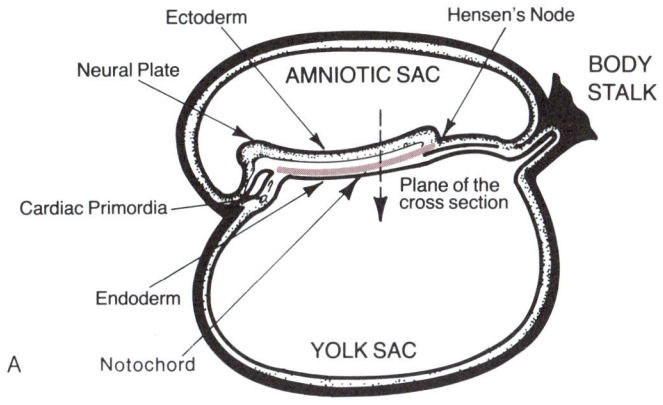

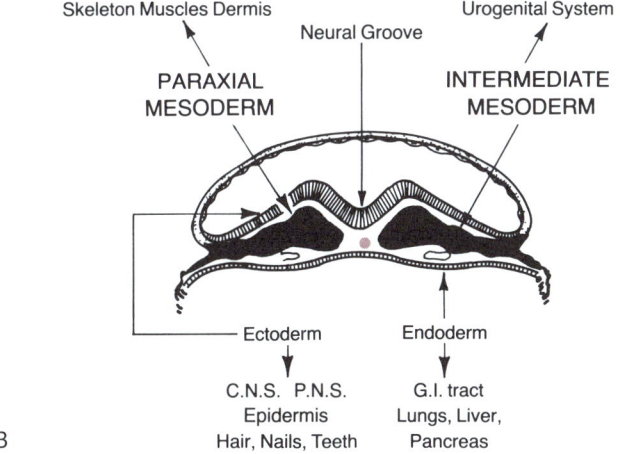

FIG. 3.1. A: Sagittal section of the presomite human embryo at 19 days. The trilaminar disk, consisting of ectoderm, mesoderm, and endoderm, is suspended between the amniotic sac and the yoke sac. **B:** Cross section of the trilaminar disk taken at the position indicated by broken line at the top. The central notochord is the embryonic axis, organizing the mesoderm and inducing neurulation. Adapted from Crelin (1981).

modistal direction; the upper extremities appear and develop earlier than the lower extremities, and the shoulder develops before the wrist.

THE EMBRYONIC PERIOD

The fourth to the eighth week is the embryonic period, a critical time when the three germ layers give rise to all the organ systems. Both the axial and the appendicular cartilaginous skeletons form at this time. The embryo is

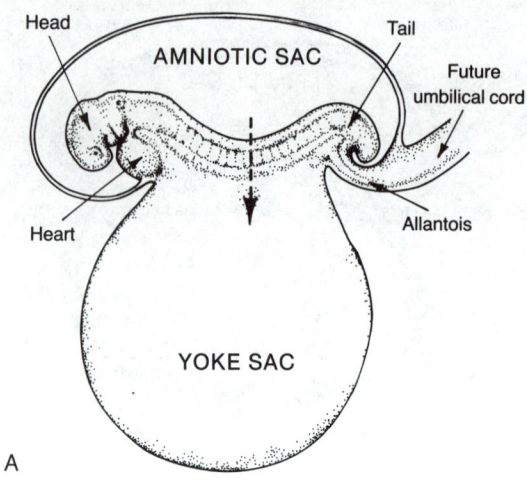

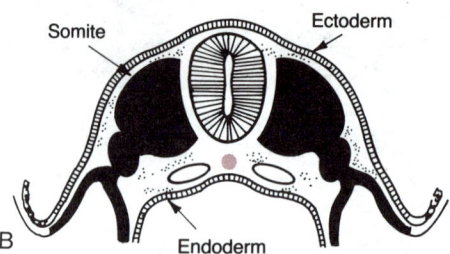

FIG. 3.2. A: A 25-day-old embryo, showing the lateral bulges of the somites, which appear in a cranial to caudal direction. (Plane of the cross section is indicated on the figure.) **B:** Cross section of the embryo. The somites are directly below the ectoderm, next to the neural tube.

particularly susceptible to malformations if exposed to viruses or teratogens during the embryonic period.

Concept 5: Most nongenetic congenital malformations occur from an insult during the first trimester. The period of greatest sensitivity is the second month of gestation, the time of organogenesis.

AXIAL SKELETON DEVELOPMENT

The precartilaginous spine forms during the fourth week by a rearrangement of sclerotomes. The sclerotomes are considered metameric structures; i.e., they exist as a series of paired homologous segments in the embryo. At

EMBRYOLOGY AND GROWTH

about the same time the spine is forming, the heart and kidneys are developing, and the trachea and esophagus begin to differentiate (a noxious influence during this time produces the VACTER association).

Concept 6: Two adjacent sclerotomes and the intersegmental mesenchyme rearrange to form a vertebra. The relationship of the spine to arteries and nerves is a result of this caudal–cranial sclerotomic resegmentation.

Spinal formation begins when cells of the sclerotome migrate medially from each somite to surround the notochord and the neural tube (Fig. 3.3). These sclerotomes are visible as a dense collection of cells at each segment;

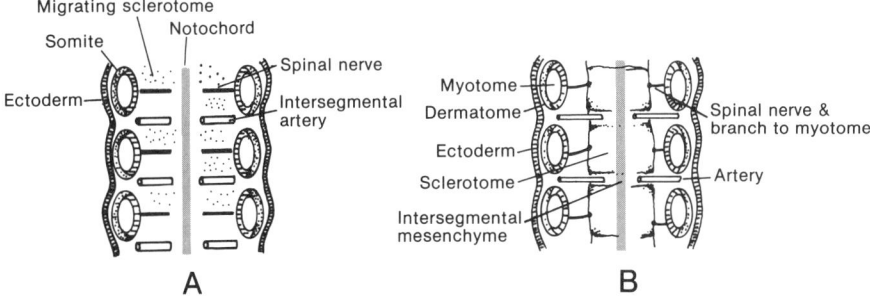

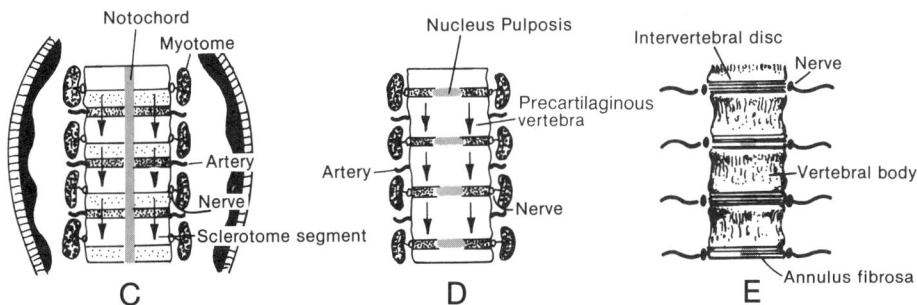

FIG. 3.3. **A:** Early in the fourth week, cells migrate medially from the somite to surround the notochord. Nerves grow laterally, directly into each somite. **B:** The notochord organizes the migrating cells into spinal sclerotomes and intersegmental mesenchyme. The somite has differentiated into the myotome and the dermatome. **C:** Cells in the caudal half of one sclerotome migrate into the intersegmental mesenchyme and cranial half of the sclerotome below. This rearrangement abolishes the intersegmental mesenchyme, and the central area of the initial spinal sclerotome becomes the intervertebral disk. **D:** Segmental arteries now enter the center of the precartilaginous vertebral bodies, while the nerves exit between the bodies. The notochord remains as the nucleus pulposus. **E:** Mature spinal configuration: each vertebral body came from the upper and lower halves of two adjacent sclerotomes.

each sclerotome is separated by a less dense area called the intersegmental mesenchyme (mesenchyme means embryonic connective tissue). Subsequently, cells from the caudal half of one sclerotome (a hemimetamere) proliferate and migrate into the intersegmental mesenchyme and into the cells of the cranial half of the subjacent sclerotome, a process called **sclerotomic resegmentation.** After resegmentation, the precartilaginous vertebral body develops from the rearranged mesenchyme, and cells previously in the center of the original sclerotome form the intervertebral disk.

As the spine matures, the notocord regresses and disappears inside the cartilaginous vertebral bodies, only remaining as the nucleus pulposis between the bodies. Occasionally, notochordal remnants persist in the basisphenoid or coccygeal region of some adults and give rise to a chordoma.

Dorsal mesenchyme from the sclerotomes covers the neural tube and eventually makes the vertebral arches while the ventral mesenchyme of the sclerotome is rearranging into the vertebral bodies. Since the segmental arteries initially grew between successive sclerotomes, they come to lie in the center of each vertebral body after the caudal–cranial resegmentation. On the other hand, the nerves originally grew into the center of each somite, so after the resegmentation, they come to lie between successive vertebral bodies and exit the canal through neural foramen.

Three primary centers of vertebral ossification appear at the end of the embryonic period. One is in the central mass of the vertebral body (the centrum), and one is in each half of the vertebral arches. At birth, the three ossification centers are connected by cartilage, usually fusing with the centrum by the age of 3 to 6 years (Fig. 3.4). Five secondary centers of ossification appear after puberty: one on the tip of the spinous process, one on the tip of each transverse process, and two ring apophyses (on the top and bottom surfaces of each body). All the centers usually unite by the age of 25 years.

The embryology of the **atlas** and **axis** is different from that of the rest of the spine. The caudal half of the last occipital sclerotome plus the first cervical and the cranial half of the second cervical combine to generate both the odontoid process and the anterior arch of the atlas. The centrum of the axis comes from sclerotome of the second and third cervical somites. Resegmentation is incomplete in the second cervical sclerotome, which results in fusion of the odontoid to the axis.

The most common spinal malformation is **spina bifida occulta,** occurring in about 20 percent of the population. This happens when the posterior borders of the neural arch fail to fuse, but it rarely produces neurological damage. However, a more serious defect occurs if the posterior neural tube fails to close as it should by the 28th day. This causes **myelomeningocele,** in which the arch and covering soft tissue fail to form and neural tissue is exposed. Almost all of these children have some neurological deficit, the severity depending on the size and location of the defect. Although the etiol-

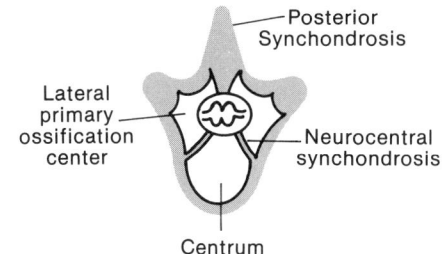

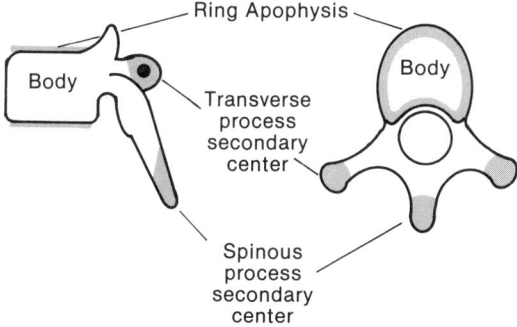

FIG. 3.4. A typical vertebra showing the three ossification centers present at birth. The neurocentral synchondrosis separates the centrum from the lateral centers, which are separated by the posterior synchondrosis.

ogy remains obscure, *in utero* detection of myelomeningocele is now possible by measuring the level of α-**fetoprotein** in the mother's serum or urine.

Other congenital deformities of the spine result from either defects of formation, defects of segmentation, or combinations of the two. The **defects of formation** produce varying degrees of wedged vertebra, hemivertebra, and butterfly vertebra (Fig. 3.5). In congenital kyphosis, failure of formation of the anterior elements sharply angulates the spine as a result of either the absence of a vertebral body or a hypoplastic one. When one side of the centrum and lateral elements fail, only one half of a vertebra appears; the result is a hemivertebra, usually producing scoliosis. If the centrum fails but the lateral centers of chondrification appear, the result is a butterfly vertebra.

Defects of segmentation produce unsegmented bars, laminar synostosis, or "block" vertebrae (Fig. 3.6). A patient with congenital scoliosis may have a complex collection of hemivertebrae, butterfly vertebrae, and unsegmented bars as a result of a combination of formation and segmentation defects.

When vertebral defects occur, rib anomalies are often present as fused or

Hypoplastic vertebra

Hemivertebra

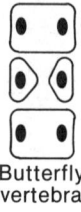
Butterfly vertebra

FIG. 3.5. Some spinal deformities resulting from defects of formation: congenital kyphosis from a hypoplastic centrum, hemivertebra, and butterfly vertebra. Adapted from Winter (1983).

absent elements. If adhesions or a fibrous stalk connect the cord or dura to the skin, the **Arnold–Chiari** deformity occurs (caudal displacement of the brainstem through the foramen magnum).

One of the more common spinal malformations is the **Klippel–Feil** syndrome, a segmentation failure of two or more cervical vertebra. Block vertebra in Klippel–Feil syndrome give these people a short neck, a low-set hairline, and a decreased range of neck motion. Thoracic and lumbar vertebra may also be involved. These people commonly have renal anomalies because the renal system is evolving from the intermediate mesoderm at the

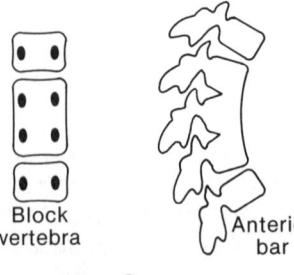
Block vertebra   Anterior bar

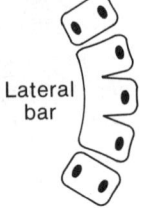
Lateral bar

FIG. 3.6. Some spinal deformities resulting from defects of segmentation: block vertebra, congenital kyphosis, and congenital scoliosis caused by an unsegmented bar. Adapted from Winter (1983).

same time and in proximity to vertebral formation from the paraxial mesoderm.

THE LIMBS

The limb buds first appear as small mounds on the ventrolateral surface of the body wall. They originate from the lateral somatic mesoderm and are covered by ectoderm; the upper limb buds appear first at 26 days, followed at 28 days by the lower limb buds. The arm buds develop opposite the fifth to the seventh cervical somites, and the leg buds, slightly larger, form opposite the lower four lumbar and first three sacral somites.

Subsequent limb development is rapid. By the fifth week, mesenchymal primordia are present, and by the seventh week, hand and foot plates are apparent. Cartilage miniatures of most of the bones-to-be are well formed by eight weeks. The clavicle is then the first bone to ossify, followed by the primary ossification center for the humerus and that of the femur during the latter part of the eighth week.

Early developmental events are the same for both upper and lower limbs except that the arms appear first, differentiate sooner, and reach their final size before the legs. In both the arm and leg buds, a structure called the **apical ectodermal ridge (AER)** promotes and controls outgrowth of the limb mesenchyme. The AER is a specialized layer of ectoderm that exerts a trophic and permissive influence on the limb mesenchyme. If the AER is surgically removed, limb formation stops, but the parts already formed continue to grow normally. Thus, once present, limb mesenchyme is a genetically determined, self-differentiating mass.

Toxic or noxious influences on the AER probably account for many limb malformations. For instance, transverse hemimelia results from a total destruction of the AER, whereas intercalary hemimelia or phocomelia results from a more localized damage to the AER. Extra tissue in the AER generates duplications such as polydactyly.

The AER directs the flipper-like limb buds to grow and eventually flatten at the ends into paddle-like hand and foot plates. An uninterrupted length of primitive mesenchyme initially occupies the core of the limbs, but within 2 weeks, the mesenchyme differentiates into cartilage models, and surface constrictions occur at the future joint sites. Cartilaginous models replace the mesoderm in a proximodistal sequence (Fig. 3.7).

Nerves grow into the limb bud beginning early in the fifth week. Cervicothoracic metamers innervate the upper limb buds (the future brachial plexus); lumbosacral metamers grow into the lower limb buds (the future lumbosacral plexus). As the limbs elongate and rotate, the nerves migrate with the limbs.

Myoblasts aggregate, elongate, and fuse into multinucleated cells. The proximal muscle masses appear before the distal, and the flexors before the

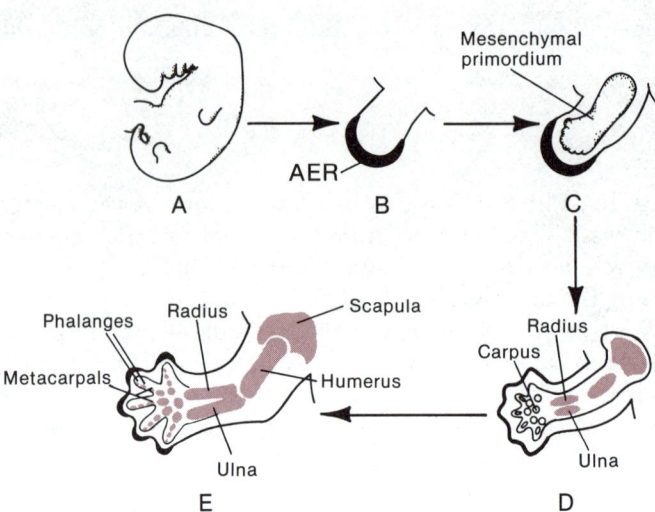

FIG. 3.7. A: The limb buds sprout from the lateral somatic mesoderm and grow laterally and caudally. **B:** Magnified view of the limb bud, showing the ectodermal covering and the terminal apical ectodermal ridge. **C:** The center of the limb is occupied by a continuous tube of mesenchyme. **D:** Centers of chondrification appear in a proximodistal gradient. **E:** By 6 weeks, cartilaginous miniatures of the bones-to-be are present. The hand plate is still in the shape of a mitten, and the AER remains only on the tips of each digit.

extensors. The radial nerve goes with the extensors, and the median and ulnar nerves with the flexors (Fig. 3.8). Muscle masses split into individual muscles by the seventh week. Nerves must be present for normal growth, and absence of an early neural influence is thought to be responsible for some disorders such as arthrogryposis multiplex congenita.

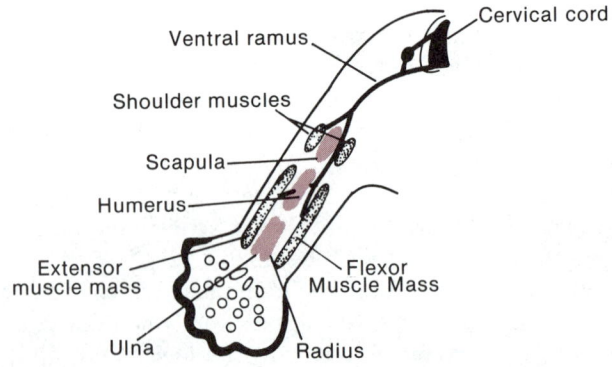

FIG. 3.8. The upper extremity at 5 weeks, showing innervation of the extensor muscle mass by the radial nerve and the flexor muscle mass by the ulnar nerve. Chondrification centers are present in the arm and forearm but have not yet appeared in the hand. By the seventh week, the muscle masses will have split into the individual muscles of the extremity. Adapted from Crelin (1981).

EMBRYOLOGY AND GROWTH

Concept 7: Final limb position results from a 90-degree external rotation of the arm buds and a 90-degree internal rotation of the leg buds.

The limb buds start out at right angles to the body wall such that the thumbs and great toes point toward the cranial (preaxial) direction. In this orientation, flexor muscles are ventral, and extensors are dorsal. Upper and lower limbs then rotate in opposite directions. The upper limbs rotate 90 degrees to put the forearm flexors medial and the extensors lateral, while the thumbs are ventral and the elbows are dorsal. The lower limbs rotate medially 90 degrees to put the extensors ventral and the flexors dorsal (Fig. 3.9). The history of lower limb rotation is reflected in the well-known spiral twisting of the hip capsule.

In addition to rotation, the upper limbs also migrate caudally, and Sprengel deformity appears if caudal migration is blocked. This can be caused by an anomalous omovertebral bone connecting a cervical vertebra to the scapula, or a similarly situated fibrous or cartilaginous band can also cause Sprengel deformity.

By the seventh week, the forearms are in pronation, and the proximal forearm mesenchymal cells begin to separate at the proximal radioulnar joints. If these cells do not separate and disappear, radioulnar synostosis occurs, most often with the forearm in pronation. Cells between the rays of the mitten-like hands and feet must also degenerate and disappear, or the result is

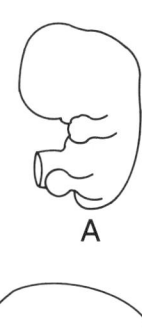

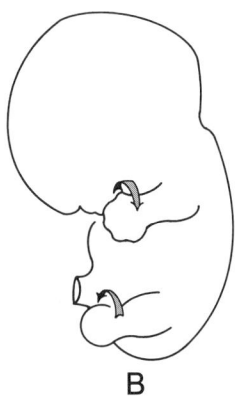

FIG. 3.9. A: At 5½ weeks, the limbs are in a preaxial orientation with the thumbs and great toes pointing cephalid. **B:** At 7 weeks, the limbs rotate 90 degrees about their long axis, but in opposite directions; the upper limbs rotate laterally, the lower limbs rotate medially.

syndactyly. Most frequently, syndactyly occurs between the long and ring fingers and the second and third toes.

JOINTS

While the AER is directing the outgrowth of the limb, the future bones appear as a continuous length of scleroblastema (primative skeletal mesenchyme). During the sixth week, sites of the future joints begin as an area called the **interzone.** Hereditary factors seem to determine the initial joint architecture since the basic form appears before muscular activity. Three events are characteristic of joint formation: (1) interzone formation, (2) synovial mesenchyme differentiation, and (3) cavitation.

Concept 8: Basic joint morphology is genetically determined, but normal joint development requires motion.

The interzone develops as a disk of densely packed cells between the scleroblastema. Homogeneous cells of the interzone then differentiate into three

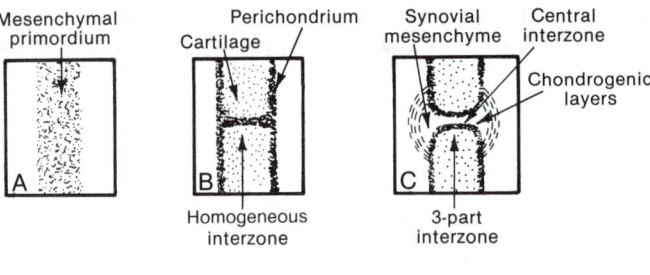

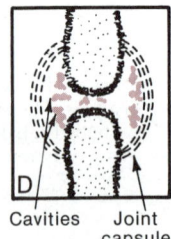
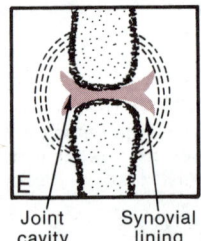

FIG. 3.10. A: The early limb bud consisting of a central core of scleroblastema (mesenchymal primordium). **B:** Appearance of the interzone, site of the future joint development. The scleroblastema has differentiated into cartilage. **C:** Three-part interzone consisting of the central interzone, the synovial mesenchyme, and the chondrogenic layers. **D:** Multiple cavities appear in the central interzone as the joint capsule develops from the synovial mesenchyme. **E:** The multiple cavities coalesce into one joint cavity characteristic of the mature joint. Reprinted with permission from O'Rahilly and Gardner (1978).

parts, the chondrogenic layers next to the scleroblastema, the synovial mesenchyme, and the central interzone.

Articular cartilage originates from the chondrogenic layers. The synovial mesenchyme gives rise to the synovial lining, the joint capsule, the menisci, and the intracapsular ligaments. The outer layer of the joint capsule becomes fibrous and maintains contact with the perichondrium of the cartilaginous bones. The inner layer becomes the highly vascularized synovial membranes.

Cells in the central interzone regress, and small cavities begin to appear as the extracellular matrix dissolves (enzymatic events responsible for cavitation remain obscure). The small spaces rapidly coalesce, forming one large joint cavity containing joint fluid (Fig. 3.10). Soon after cavitation, motion begins from the maturation of innervated muscle. Not only is motion necessary to help shape the joint surfaces, but it is essential for complete joint development.

The knee joint goes through an intermediate stage in which it is divided into three chambers by walls of the synovial mesenchyme. If these walls fail to disappear entirely, the membranes remain as **synovial plica,** the most common being the infrapatellar plica (ligamentum mucosum), then the suprapatellar plica, and, least common (but most symptomatic), the medial plica.

Other anomalies of cavitation can occur as a result of incomplete interzone formation. For instance, a persisting bridge of scleroblastema can traverse the interzone, producing bony coalitions such as the tarsal and carpal coalitions. As mentioned in Chapter 1, tarsal and carpal coalitions have a genetic basis and are inherited as autosomal dominant traits.

BONE FORMATION

Each of the more than 200 bones of the body has a unique developmental history, but in all cases, bone never forms *de novo;* it is always made in an area occupied by either fibrous or cartilaginous tissue. Intramembranous bone comes from fibrous-tissue derivatives and includes the skull and facial bones, parts of the clavicle and mandible, and all subperiosteal bone. The remainder of the skeleton comes from cartilaginous tissue derivatives by the process of endochondral bone formation. Although intramembranous and endochondral bone formation are different histological processes, the type of bone made is the same, namely, woven bone. It is the environment in which bone is made that is different, not the final bone product.

Intramembranous bone formation involves osteoblasts differentiating directly from the mesenchyme, lining up on a fibrous tissue strut, and secreting a layer of osteoid that mineralizes. The fibrous strand is called a trabecula (L. little beam) after mineralization. Some osteoblasts on the trabeculae are trapped in the osteoid layer and become osteocytes. The surface osteoblasts continue to lay down more osteoid. Individual shafts of trabeculae join to produce a three-dimensional lattice framework of primary cancellous bone that takes a genetically predetermined shape of the clavicle, the temporal bone, etc., as the case may be.

Intramembranous ossification is also responsible for growth in width of all the long bones. Growth in length is accomplished by endochondral ossification. The first bone made is woven bone, having a random orientation of collagen fibers. This woven bone is eventually resorbed and replaced by much stronger lamellar bone, in which the collagen fibers are arranged in orderly arrays within each lamella.

A preexisting cartilage model is required for **endochondral bone formation** (Fig. 3.11). The cartilage is not simply converted into bone; it is mostly destroyed, and bone then forms on the remnants of the extracellular matrix.

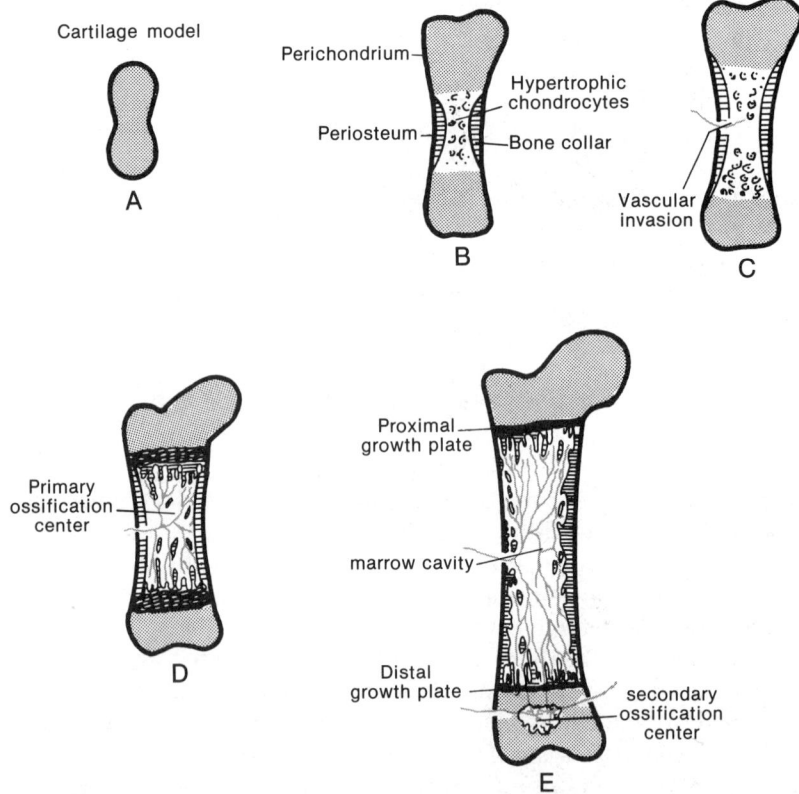

FIG. 3.11. Stages in bone maturation. **A:** The bone-to-be starts out as a cartilage model. **B:** Central chondrocytes hypertrophy as a collar of intramembranous bone encircles the diaphysis. **C:** The nutrient vessel penetrates the bone collar and invades the degenerating chondrocytes, **(D)** establishing the primary center of ossification. **E:** The secondary ossification center appears in the epiphysis; note the transphyseal vessels.

EMBRYOLOGY AND GROWTH

All the cartilaginous miniatures of the long bones share a similar fate. Osteoblasts make a bone collar surrounding the midshaft, and the central chondrocytes hypertrophy and degenerate. Vascular tissue penetrates the bone collar to invade the hypertrophic, degenerating chondrocytes. Osteoblasts follow the vascular penetration and form the primary center of ossification that grows toward both ends of the cartilage model. Later, a similar vascular invasion of the epiphyses starts the secondary center of ossification. The disk of growing cartilage separating the epiphysis from metaphysis is the growth plate, or physis (Gr. to generate). Endochondral ossification occurs at both the epiphyseal and metaphyseal surfaces of the growth plate, but it is much faster on the metaphyseal side. More specific details of the growth plate are discussed below.

THE FETAL PERIOD

At 9 weeks, the embryo is considered a fetus (L. offspring), and it has a recognizable human appearance. Major tissues and organ systems have formed, and the next 7 months are concerned mostly with growth and differentiation (Table 3.1). A fetus is much less vulnerable to drugs and viruses than an embryo.

The body grows in length and width, and by 12 weeks the upper limbs reach their normal relative proportions. By 20 weeks, the legs reach their

TABLE 3.1. *Summary of the major embryonic events*

Day	Embryonic event
1	Zygote
3	Morula
5	Blastocyst
10	Implantation
20	Primitive streak
21	Complete notochord
22	Neural groove
24	Somites
26	Arm buds
28	Leg buds
29	Elbow crease
33	Hand plate
37	Foot plate
38	Cartilage miniatures
41	Elbow bends
42	Finger rays
44	Toe rays
45	Clavicle ossified
48	Humerus primary ossification center
52	Adult external appearance

TABLE 3.2. *The prevalence of some musculoskeletal abnormalities*[a]

	Rate per 1,000 births
Myelomeningocele	0.2
Congenital hip dislocation	1.8
Syndactyly	2.5
Clubfoot	3.8
Polydactyly	7.4

[a]From Graxier et al. (1984).

final relative proportions, and the mother can detect fetal movements. After 26 to 29 weeks, a fetus can survive premature delivery with intensive neonatal care, but the mortality rate is high, as is the incidence of disorders such as cerebral palsy.

Most full-term infants are normal, but 2 to 3 percent are noted to have anatomical abnormalities at birth (Table 3.2). Recall from Chapter 1 that congenital malformations can be caused by either genetic or environmental factors. The major genetic factors include (1) numerical chromosomal disorders such as monosomy or trisomy, (2) structural chromosomal disorders such as chromosome breaks and duplications, and (3) mutations and transmissible genetic abnormalities with structurally normal chromosomes. Some of the known environmental factors include (1) viruses such as rubella, cytomegalovirus, and herpes simplex, (2) drugs such as phenytoin, warfarin, thalidomide, and alcohol, and (3) ionizing radiation.

FUNCTIONAL ANATOMY OF THE GROWTH PLATE

The growth plate can be divided into three parts for the sake of discussion. The first part is the **growth cartilage,** divided into different morphologic zones. The next part is the **metaphysis,** where bone is made on the remnants of calcified cartilage. The last part is the circumferential fibrous structure consisting of the perichondrial **ring of LaCroix** and the **groove of Ranvier.**

Chondrocytes in the growth cartilage go through a series of cellular changes producing longitudinal bone growth. These changes produce three different histologic zones, the **reserve zone,** the **proliferating zone,** and the **hypertrophic zone.** The hypertrophic zone is subdivided further into three more zones, the zone of **maturation,** the zone of **degeneration,** and the zone of **provisional calcification.** Cell physiology at each of the zones depends to a large extent on the local blood supply and resulting oxygen tension.

Three major arterial systems supply the growth plate: the epiphyseal

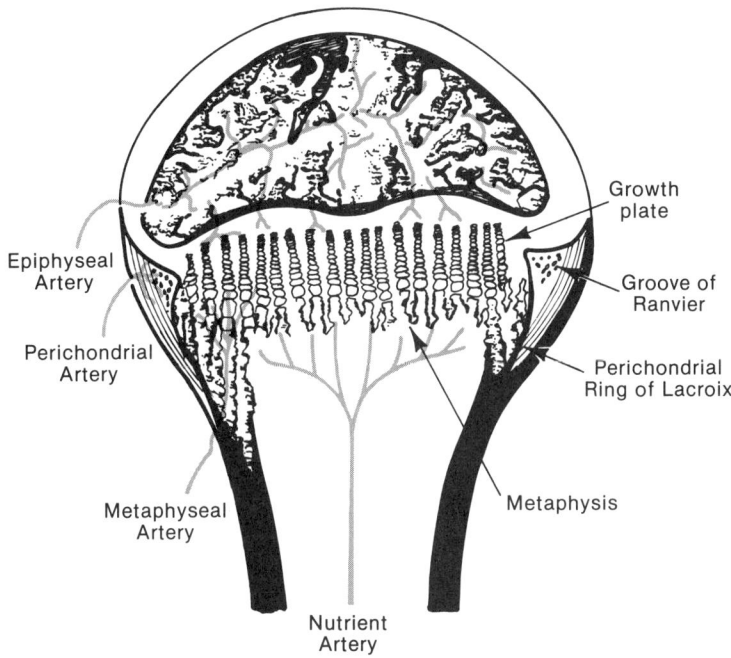

FIG. 3.12. Schematic drawing of the growth plate and blood supply. The epiphyseal artery supplies the secondary ossification center and provides circulation to the growth cartilage. The perichondrial artery supplies the groove of Ranvier and LaCroix's ring. Vessels from the metaphyseal and nutrient arteries grow into the zone of provisional calcification.

artery, the nutrient artery, and the perichondral circulation (Fig. 3.12). The **epiphyseal artery** sends branches through cartilage canals to supply the growth cartilage cells, but capillaries penetrate only to the level of the proliferative zone. The hypertrophic zone is avascular.

Branches of the **nutrient artery** (and also the metaphyseal vessels) supply a rich vascular network to the metaphysis. Closed capillary loops grow into the vacated hypertrophic lacunae as the transverse septae are dissolved.

The last arterial system, the **perichondrial arteries,** supply the ring of LaCroix and the groove of Ranvier. Capillaries from this system communicate with the epiphyseal and metaphyseal capillaries as well as the joint capsule vessels and can be a potential route of extension of metaphyseal osteomyelitis into the epiphysis or into the joint.

Morphology of the Growth Cartilage

The **reserve zone** contains a sparse collection of chondrocytes widely separated by extracellular matrix. These chondrocytes have abundant rough endoplasmic reticulum and numerous lipid and glycogen granules, which reflect their active metabolic role of matrix synthesis, but the calcium content is low, as is the local oxygen tension (20 mm Hg). Arteries and veins traverse this zone in special conduits called cartilage canals on the way to the proliferative zone; little oxygen is unloaded through the canals. Functionally, the reserve zone provides a continuous supply of chondrocytes to the proliferative zone.

The chondrocytes flatten and undergo mitotic division in the **proliferative zone.** As the cells divide, they line up in longitudinal columns parallel to the long axis of the bone. Each column is surrounded by thick longitudinal partitions (septae) of cartilage, and cells within each column are close to each other, separated by a thin transverse cartilage septum. Matrix constitutes 75 percent of the volume of the proliferative zone, and cells 25 percent. The capillary network is rich around the proliferating chondrocytes. Oxygen tension is high (57 mm Hg), and chondrocytes use aerobic metabolism to synthesize and store glycogen. Most growth in length of the bones results from the division and interstitial growth of cells in this zone. It is important to understand that cells do not actually move within the zone, but the epiphysis moves away as a result of cell division.

Chondrocytes in the zone of **hypertrophy** enlarge to five times their original size and ultimately degenerate and die at the bottom of the zone. Enlargement is at the expense of extracellular matrix, and both longitudinal and transverse septae become thinner. Cells now occupy about 60 percent of the volume of the zone, and matrix about 40 percent. Morphological and biochemical changes can be used to divide the hypertrophic zone into three subzones.

The top portion is the zone of **maturation,** where proliferative cells begin to enlarge and consume their glycogen stores. Mitochondria, which have the capacity to store ions, become loaded with calcium, and the extracellular matrix contains normal, highly aggregated proteoglycans.

The middle portion is the zone of **degeneration,** where cells begin to show signs of intracellular deterioration. Oxygen tension is low (24 mm Hg), and the cells shift to anaerobic metabolism, using the glycogen stores for substrate. As the glycogen energy source is depleted, subcellular organelles fragment, and the cytoplasm becomes vacuolated. Mitochondria begin to release the stored calcium, and matrix proteoglycans disaggregate.

At the bottom portion, the zone of **provisional calcification,** chondrocytes are either moribund or dead, and calcium salts diffuse into and deposit in the longitudinal septae. The proteoglycans have disaggregated (perhaps rearranging into rosette-like structures that organize the calcification). Crystals appear in matrix vesicles. The growth cartilage ends, and the metaphysis begins below the last intact transverse septae (Fig. 3.13). Inhibition of cartilage matrix calcification, as occurs in rickets, prevents normal vascular invasion through the transverse septae.

EMBRYOLOGY AND GROWTH

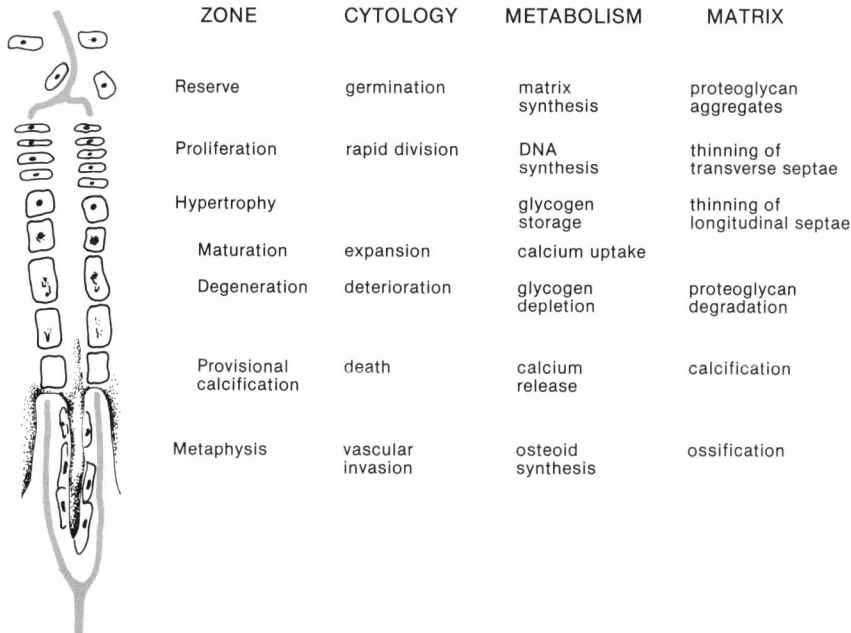

FIG. 3.13. Schematic representation of the various zones of the growth plate with the corresponding cellular, metabolic, and matrix events.

Metaphysis

Bone formation and bone remodeling are the major events in the metaphysis. Bone formation begins with dissolution of the uncalcified transverse septae of the hypertrophic zone followed by capillary endothelial invasion of the hypertrophic lacunae. The vascular sprouts enter a barren area of honeycombed lacunae, devoid of living chondrocytes, low in oxygen tension (19 mm Hg), and confined by the calcified longitudinal septae (Fig. 3.14).

Calcified cartilage spicules extending from the growth plate into the metaphysis are called primary spongiosum. As the capillaries advance, perivascular osteoblasts adhere to the primary spongiosum and begin synthesizing osteoid, which eventually mineralizes. This combination of calcified cartilage septum covered by layers of new bone is called secondary spongiosum (Fig. 3.15).

Remodeling begins immediately. Osteoclasts resorb the secondary spongiosum, and osteoblasts closely follow, making lamellar bone in the wake of the osteoclasts. In this way, the initial woven bone is completely replaced by lamellar bone. Osteoclasts also resorb, or cut back, the peripheral bone of the metaphysis. This remodeling gradually narrows the width of the metaphysis down to that of the diaphysis.

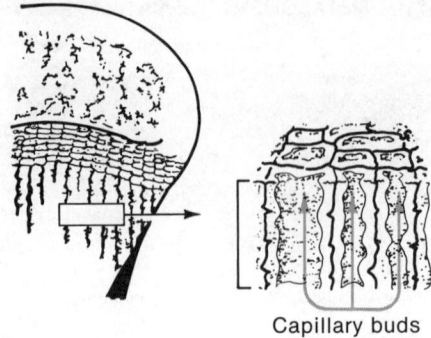

FIG. 3.14. The metaphyseal honeycomb of vascular invasion. The area is devoid of viable chondrocytes, has a low oxygen tension, and is confined by calcified cartilaginous walls of the longitudinal septae. Adapted with permission from Ham (1969).

Circumferential Structure

The perichondrial **ring of LaCroix** and the ossification **groove of Ranvier** surround the growth plate. Both the ring and the groove are part of the same anatomical structure, but they have different biological functions. The perichondrial ring, studied in detail by LaCroix, is the peripheral supporting girdle of the growth plate. It contains circumferential collagen fibers, forming a fibrous collar around the growth cartilage and the metaphysis. The ring is continuous with the fibrous layer of the periosteum. In some growth plates,

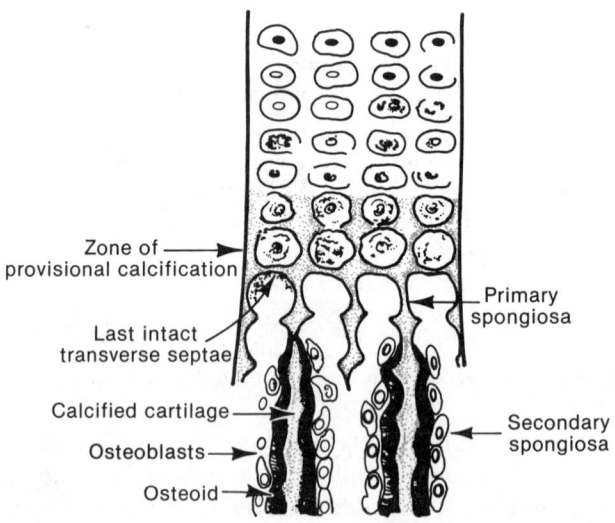

FIG. 3.15. Primary spongiosa are the calcified cartilage bars extending into the metaphysis that are the remnants of the longitudinal septae. Secondary spongiosa are formed by the addition of layers of woven bone on the primary spongiosa. Adapted from Ham (1952).

the inner layer of the ring is lined by a thin sheath of woven bone, called the "bone bark" by Ranvier.

The ossification groove is a wedge-shaped collection of cells jutting out into the plate in the region of the reserve and proliferative zones. The groove is an area of intense cell division, replenishing the perichondrial ring cells and providing cells for growth in diameter of the plate. Together, the perichondrial ring and ossification groove support and expand the growth plate.

FURTHER GROWTH OF BONE (MODELING)

Bones grow not only in length but also in width and thickness. As described above, the morphological events in the growth cartilage increase the length of a bone. However, the diameter of the metaphysis is larger than the diaphysis, so the metaphysis must be cut back to the width of the diaphysis. This modeling process is termed **funnelization** and is accomplished by a collection of osteoclasts and osteoblasts along the periphery of the metaphysis. The continual activity of these cells is responsible for the thin, fenestrated proximal metaphyseal surface.

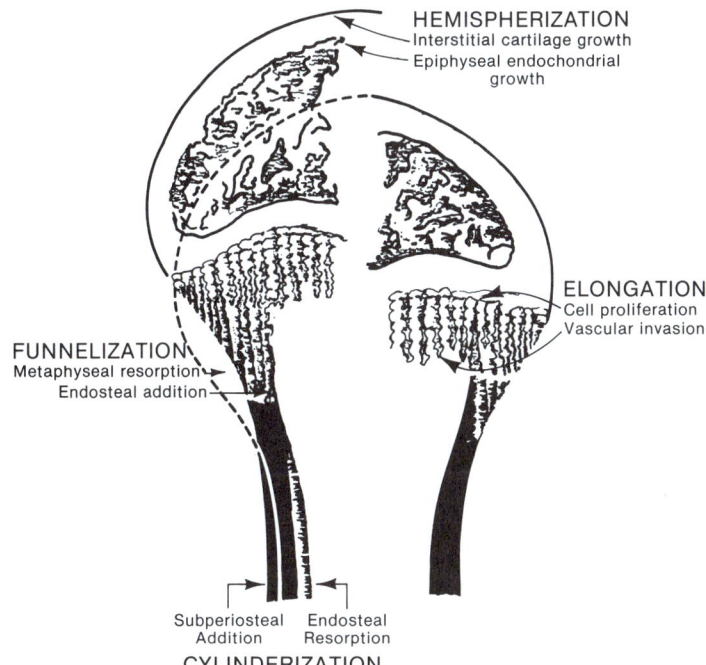

FIG. 3.16. Remodeling of growing bone from selective resorption and deposition of bone in the epiphysis, metaphysis, and diaphysis, accounting for bone growth and change in shape. Adapted from Ham (1952).

The width of bones increases as a result of intramembranous appositional bone formation by the periosteum. As new layers are laid down on the outside, osteoclasts resorb older bone at the endosteal surface on the inside, increasing the medullary cavity diameter. The rate of bone formation on the outside is slightly greater than the rate of resorption on the inside, so the cortex grows in thickness as well as in diameter. This modeling process is called **cylinderization.**

A modification of endochondral ossification is also responsible for growth of the epiphysis (the end of the bone) by the process of **hemispherization.** This is best understood by recalling that the ends of a bone are covered by two types of cartilage, a thin layer of growth cartilage directly below a thicker layer of articular cartilage. Endochondral ossification progresses in this growth cartilage layer, enlarging the hemisphere of the epiphysis, but at a slower rate and with less formal organization than in the growth plate (physis). Thus, a bone acquires its final shape by the coordination of growth plate endochondral ossification with the modeling processes of funnelization, cylinderization, and hemispherization (Fig. 3.16).

GLOSSARY

Anomaly A marked deviation from the anatomically normal standard as a result of congenital or hereditary defects.

Apical ectodermal ridge The AER is the cap of ectoderm covering the end of the limb bud that is responsible for orderly growth of the bud.

Blastocyst A hollow ball of cells, filled with fluid, that appears 4 days after conception.

Cylinderization The remodeling process creating the cylindrical diaphysis of a bone, where periosteal new bone is made as osteoclasts resorb endosteal bone to widen the medullary canal.

Dermatome Lateral portion of the mesodermal somite giving rise to segmental subcutaneous tissue.

Diaphysis Shaft of the bone, formed from the primary center of ossification, extending from the proximal to the distal metaphysis.

Ectoderm The outer layer of the three primary germ layers, giving rise to the skin, nails, hair, glands, nervous system, ears, and eyes.

Endoderm The inner layer of the three primary germ layers, giving rise to the epithelium of the respiratory, gastrointestinal, and urogenital systems, also called the entoderm.

Epiphysis The end of a long bone, usually formed from a secondary ossification center and separated from the shaft of a growing bone by the growth plate.

Funnelization The remodeling process in which the metaphysis is cut back to the diameter of the diaphysis as a long bone grows.

Hemispherization The growth and modeling process enlarging and expanding the epiphyseal end of a bone.

Metameric Developing structures that are part of homologous segments of the body, such as sclerotomes.

Metaphysis Flared part of the bone between the shaft and epiphysis. That region of a growing bone from the bottom of the growth plate to the diaphysis.

Morula Solid ball of cells formed by 3 days from the dividing fertilized ovum.

Myotome Part of a somite that develops into skeletal muscle.

Notochord Rod-shaped collection of mesodermal cells, below the primative groove of the ectoderm, forming the axis of the embryo, around which the spine and nervous system develop.

Ossification groove of Ranvier A groove of cells circling the growth cartilage providing cells for growth in diameter of the growth plate.

Osteoid The extracellular organic matrix of immature bone, made by osteoblasts, that has not yet undergone calcification.

Perichondrial ring of LaCroix A circumferential fibrous collar around the growth cartilage providing support and anchoring the epiphysis to the metaphysis.

Primary spongiosum Calcified longitudinal trabeculae at the bottom of the growth cartilage, devoid of cells and extending into the metaphysis.

Sclerotome Medial portion of the somite, giving rise to the axial skeleton.

Sclerotomic rearrangement The process of vertebral body formation in which cells in the caudal half of one sclerotome migrate into the cells of the cranial half of the sclerotome below.

Secondary spongiosum Bone containing a central core of calcified cartilage covered by woven bone.

Somites Paired masses of mesoderm, arranged alongside the neural tube, that separate into the sclerotome, myotome, and dermatome.

Zygote The cell resulting from the combination of the sperm and the egg; the fertilized ovum.

BIBLIOGRAPHY

Brighton, C. T. (1974): Clinical problems in epiphyseal plate growth and development. *A.A.O.S. Instruct. Course Lect.*, 23:108–122.

Brighton, C. T. (1984): The growth plate. *Orthop. Clin. North Am.*, 15:571–595.

Corliss, C. E. (1976): Patten's human embryology. In: *Elements of Clinical Development*, pp. 168–197. McGraw-Hill, New York.

Crelin, E. S. (1981): Development of the musculoskeletal system. *Ciba Clin. Symp.*, 33:1–36.

Cruess, R. L. (1982): Growth and its control, including the epiphysis. In: *The Musculoskeletal System. Embryology, Biochemistry, and Physiology*, edited by R. L. Cruess, pp. 191–218. Churchill Livingstone, New York.

Gardner, E. (1972): Osteogenesis in the human embryo and fetus. In: *The Biochemistry and*

Physiology of Bone, Vol. 3, edited by G. H. Bourne, pp. 77–116. Academic Press, New York.
Graxier, K. L., Holbrook, T. L., Kelsey, J. L., and Stauffer, R. N. (1984): *The Frequency of Occurrence, Impact, and Cost of Musculoskeletal Conditions in the United States,* p. 35. American Academy of Orthopedic Surgeons, Chicago.
Ham, A. W. (1952): Some histophysiological problems peculiar to calcified tissue. J. Bone Joint Surg. *34A* 701–709.
Ham, A. W. (1969): *Histology,* p. 426. J. B. Lippincott, Philadelphia.
Harrison, R. G. (1973): *Clinical Embryology.* Academic Press, New York.
Langman, J. (1981): *Medical Embryology,* fourth ed. Williams & Wilkins, Baltimore.
Ogden, J. A. (1980): Chondro-osseous development and growth. In: *Fundamental and Clinical Bone Physiology,* edited by M. R. Urist, pp. 108–171. J. B. Lippincott, Philadelphia.
O'Rahilly, R., and Gardner, E. (1978): The embryology of movable joints. In: *The Joints and Synovial Fluid,* edited by L. Sokoloff, pp. 49–103. Academic Press, New York.
Rana, M. W. (1984): *Key Facts in Embryology.* Churchill Livingstone, New York.
Shapiro, F., Holtrop, M. E., and Glimcher, M. J. (1977): Organization and cellular biology of the perichondrial ossification groove of Ranvier. *J. Bone Joint Surg.,* 54A:703–723.
Speer, D. P. (1982): Collagenous architecture of the growth plate and perichondrial ossification groove. *J. Bone Joint Surg.,* 64A:399–407.
Warshawsky, H. (1982): Embryology and development of the skeletal system. In: *The Musculoskeletal System. Embryology, Biochemistry, and Physiology,* edited by R. L. Cruess, pp. 33–56. Churchill Livingstone, New York.
Winter, R. B. (1983): *Congenital Deformities of the Spine.* Thieme-Stratton, New York.

4
Structural Components

THE EXTRACELLULAR MATRIX

Bones, ligaments, tendons, and cartilage are classified as connective tissues; they provide mechanical support, transmit forces, and maintain structural integrity of the body. All connective tissues are composed of two basic parts, cells and extracellular matrix. The cells synthesize and secrete structural components, which are assembled into an organized meshwork constituting the extracellular matrix. Although the cells are different, it is the composition and construction of the extracellular matrix that best characterizes the connective tissues.

For the most part, structural components of the extracellular matrix are macromolecules, large compounds with molecular weights often exceeding 100,000. The relative amounts of macromolecules and their organization vary considerably in connective tissues, providing a diversity that satisfies the functional requirements of the tissue. For instance, collagen in the Achilles tendon is more highly cross linked than that in the extensor digitorum tendons, and articular cartilage contains an entirely different type of collagen from that in tendons.

The extracellular matrix is not an inert or indifferent surrounding. It is a dynamic milieu, directly influencing the cells embedded within by limiting substrate diffusion, dictating cellular configuration, and restricting or promoting migration. In turn, different cells exploit the physical properties of the structural components to meet mechanical demands placed on the tissue, resisting tension in ligaments or providing rigidity in bone.

This chapter reviews the structural components of the extracellular matrix, namely, collagen, proteoglycans, elastin, fibronectin, other proteins, and, finally, mineral.

COLLAGEN: THE MAJOR STRUCTURAL MACROMOLECULE

Collagen is the most abundant protein in the body, accounting for over 30 percent of all proteins. To put this in perspective, consider that collagen is 6 percent of man's total body weight (but recall that 70 percent of man's total weight is water). Collagen constitutes 90 percent of the dry weight of osteoid, the organic matrix of bone, and about 70 percent of the dry weight of ligaments and tendons. Commercially, collagen is extracted from bones to make glue, hence the name (Gr. *kolla*, glue).

Structure of Collagen

Collagen is actually a family of proteins sharing unique amino acids and several structural features. The central feature of all collagens is a stiff helical structure, making a long, insoluble molecule with great tensile strength.

Collagen polypeptide chains contain 1,055 amino acids, in which every third amino acid is glycine, and they are also rich in proline and hydroxyproline (Fig. 4.1). The amino acid sequence actually consists of repeating units of three; every first position is glycine, and every third position is proline or hydroxyproline. The second position shows a high variation. These polypeptide chains are called α chains. Each α chain is twisted into a left-handed helix, and three α chains are coiled into a right-handed superhelix to form a macromolecule similar in appearance to a segment of cable called **tropocollagen** (Gr. *trope*, turn). Tropocollagen is a rigid rod made of three α chains that are twisted together, 300 nm in length and 1.5 nm in width, with a molecular weight of 285,000.

The strict requirement that every third amino acid in an α chain be glycine is now understood from steric analysis. The inside of a triple-stranded tropocollagen molecule is very crowded, and the only residue that can actually fit in the interior position where the three chains meet is the smallest amino acid, glycine. The high content of proline and hydroxyproline is also understood. These are cyclic, rigid amino acids that force the polypeptide chain into the left-handed helix and greatly limit rotation, thereby stabilizing the triple helix.

Hydroxyproline is not an amino acid directly incorporated into proteins

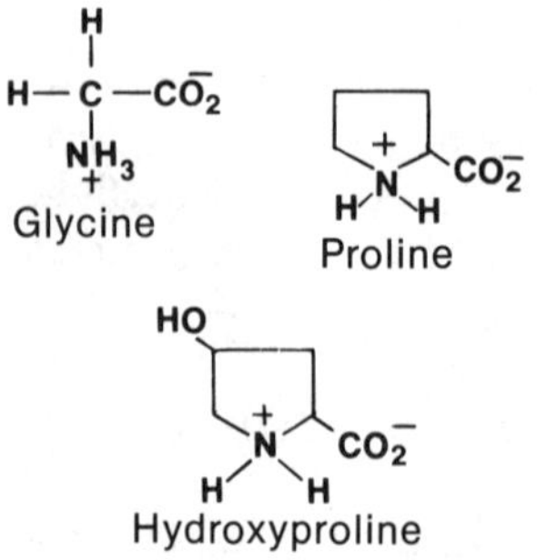

FIG. 4.1. The chemical structures of glycine, proline, and hydroxyproline. Glycine is the smallest amino acid and the only one able to fit the interior of the collagen triple helix. Proline's rigid ring structure determines the direction and pitch of the left-handed helix of the α-chain.

at the ribosome. In fact, hydroxyproline is absent from the cytoplasmic pool of amino acids, and cells lack a hydroxyproline transfer RNA. Hydroxyproline comes from the addition of a hydroxyl group to certain proline residues already in the α chain; that is, they are made after ribosomal synthesis of the polypeptide. This posttranslational hydroxylation is performed by the enzyme proline hydroxylase. Only a few other proteins are known to contain hydroxyproline (elastin, acetylcholinesterase, and the C1q subcomponent of complement). In collagen, the hydroxyl group of hydroxyproline forms intermolecular hydrogen bonds that are essential for stability of the triple helix.

Molecular Heterogeneity

Collagen molecules have a gentically determined diversity. As mentioned, all collagens are long, rigid rods of α chains containing glycine, proline, and hydroxyproline. However, the exact sequence of amino acids in the α chains and the posttranslational modifications such as proline hydroxylation and lysine glycosylation are distinct and genetically determined. These seemingly minor variations account for why bone collagen mineralizes whereas tendon collagen does not or why corneal collagen is transparent.

Ten different kinds of collagen have been identified so far, and five types are of musculoskeletal concern (Table 4.1). It is convenient to divide the five types into three categories: the **interstitial (fibrillar) collagens** type I, type II, and type III; **basal lamina collagen,** type IV; and **cell-associated collagen,** type V.

TABLE 4.1. *Types and properties of collagen*

Type	Composition	Distinctive features	Distribution
I	$[\alpha 1(I)]_2 \alpha 2(I)$	Low hydroxylysine and carbohydrate; 90 percent of all collagen; forms thick fibrils	Bones, ligaments, tendons, skin
II	$[\alpha 1(II)]_3$	High hydroxyproline and carbohydrate; forms thin fibrils	Hyaline and growth cartilage; nucleus pulposus
III	$[\alpha 1(III)]_3$	High hydroxyproline; low carbohydrate; reticulin fibrils	Blood vessels, muscle, skin, internal organs
IV	$[\alpha 1(IV)]_2 \alpha 2(IV)$	Very high hydroxylysine and carbohydrate	Basal lamina
V	$\alpha A(\alpha B)_2$	Unknown	Cell associated, widespread in small amounts

The Interstitial (Fibrillar) Collagens

Type I collagen is the most abundant, accounting for 90 percent of all collagens, and it is present in bones, tendons, ligaments, and skin. The triple helix of type I consists of two identical α chains called $\alpha 1(I)$ and one slightly different chain called $\alpha 2(I)$. Type I collagens have a propensity to polymerize into fibrils by a process called fibrillogenesis (discussed later). Bone collagen differs from skin, ligament, and tendon collagen by the presence of more intermolecular bonds.

Type II collagen is the predominant species in hyaline cartilage, growth cartilage, and nucleus pulposis. It is a triple helix of three identical α chains called $\alpha 1(II)$; the chains have a high content of hydroxylysine and covalently attached carbohydrate. Type II collagen readily associates with and tightly adheres to proteoglycans.

Type III collagen is mostly found as an interstitial component associated with type I collagen, in arteries, muscle, lung, and skin, and is a major component of the walls of hollow gastrointestinal organs. Type III is also composed of three identical α chains, $\alpha 1(III)$. It is high in hydroxyproline, low in hydroxylysine, and readily associates with other proteins, such as elastin and type I collagen. "Reticulin" fibers are thought to be type III collagen.

Basal Lamina and Cytoskeleton Collagen

Type IV collagen, found in basement membranes, makes flat arrangements by combining with other proteins such as the protein laminin (L. *lamina,* thin plate) and fibronectin (L. *nectere,* to bind). High concentrations of type IV collagen are present in glomerular basement membranes, lens capsule, and Descemet's membrane. **Type V** collagen is found principally in vessels, lung, liver, and smooth muscle cells. It is thought to be used by cells as an extracellular extension of the cytoskeleton, probably orchestrating and regulating many of the extracellular events.

Quantitative analysis of the collagen types shows that bone contains almost 100 percent type I collagen, whereas tendons contain about 95 percent type I and 5 percent type III collagen (Fig. 4.2). Hyaline cartilage is 99 percent type II collagen with small amounts of type III and type V. Fibrocartilage, formed after injury to articular cartilage, contains type I collagen mixed in with type II, which changes the physical properties of the tissue to the detriment of joints.

Synthesis

With modern molecular genetic techniques, considerable progress has been made in the understanding of collagen genes. The gene for $\alpha 1(I)$ is on

STRUCTURAL COMPONENTS

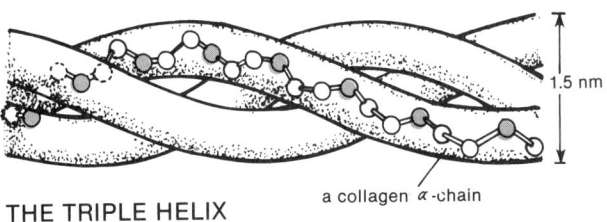

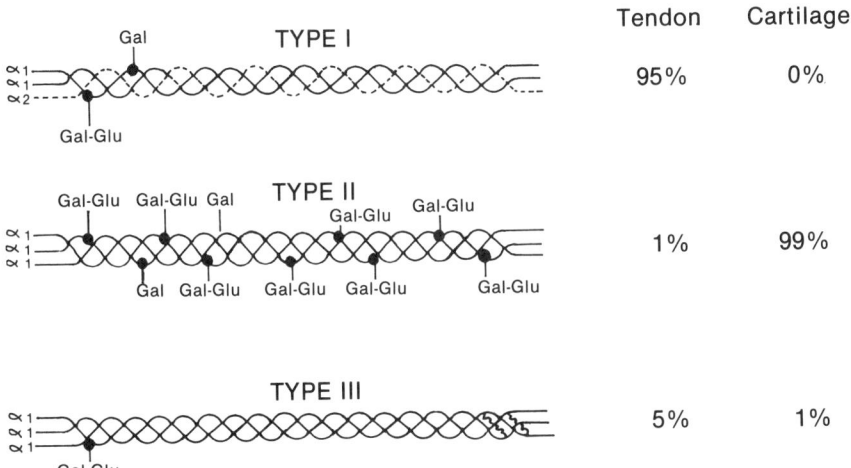

FIG. 4.2. Structure of the triple helix and schematic of the three interstitial types of collagen, showing the distribution in tendon and cartilage. There are differences in α-chain amino acid compositions collagen hydroxylation, and glycosylation among the types. Type III is the only one containing internal disulfide cross links.

the long arm of chromosome 17, and the gene for α 2(I) is on the long arm of chromosome 7 along with the gene for type IV collagen. As mentioned in Chapter 1, the α 1(I) gene is very large and contains over 50 introns (intervening sequences), ranging in size from 86 to 2,000 base pairs. After transcription, the introns are excised, and the exons (expressed sequences) are spliced into an mRNA 7,200 base pairs in length. The mature mRNA exits the nucleus for translation on membrane-bound ribosomes.

Collagen mRNA is translated only on membrane-bound ribosomes. The new polypeptides immediately pass into the endoplasmic reticulum cisternal space, never appearing free within the cytosol. This passage is directed by a short hydrophobic peptide called the "leader peptide." The leader peptide is removed immediately after the new polypeptide chain enters the cisternal space. Each new polypeptide chain (called a pro-α chain) possesses extra amino acids on the N-terminal and C-terminal ends (propeptides) that do

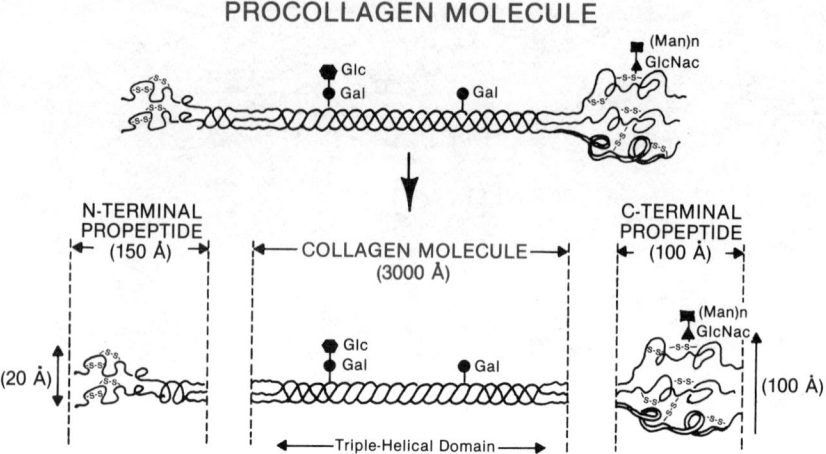

FIG. 4.3. Structure of type II procollagen molecule. After secretion into the extracellular matrix, both C-terminal and N-terminal extension peptides are removed by specific proteinases. The tropocollagen molecules self-assemble into fibrils. Glc, glucose; Gal, galactose; Man, mannose; GlcNac, N-acetylglucosamine. Adapted with permission from Prockop et al. (1979).

not appear in tropocollagen (see below). In the cisternal space, three of the appropriate pro-α chains, including the propeptides, are chemically modified by hydroxylation and glycosylation then polymerized into a triple helix called **procollagen.**

Structure and Function of Procollagen

Collagen synthesis is similar to the synthesis of other cellular proteins made for export from the cell (i.e., proinsulin, proparathyroid hormone) in that it is first synthesized as a larger precursor molecule, procollagen. The **procollagen** molecule is a helix of three polypeptides, the pro-α chains. Each pro-α chain has an N-terminal propeptide with a molecular weight of 15,000 and a C-terminal propeptide of 35,000 (Fig. 4.3). The propeptides have at least two functions: (1) they guide the intracellular formation of the triple helix (in the absence of propeptides, triple helix formation is very slow), and (2) they prevent the intracellular formation of collagen fibrils. Only after removal of the propeptides will tropocollagen aggregate into fibrils **(fibrillogenesis).**

FIG. 4.4. Intracellular and extracellular events of collagen synthesis and fibrillogenesis. Procollagen is secreted into the ER cisternae as it is synthesized and then packaged for secretion in the Golgi. Removal of propeptides permits fibril formation, aggregation, and cross linking into fibers of ligaments, tendons, etc. Adapted with permission from Albers, et al. (1983).

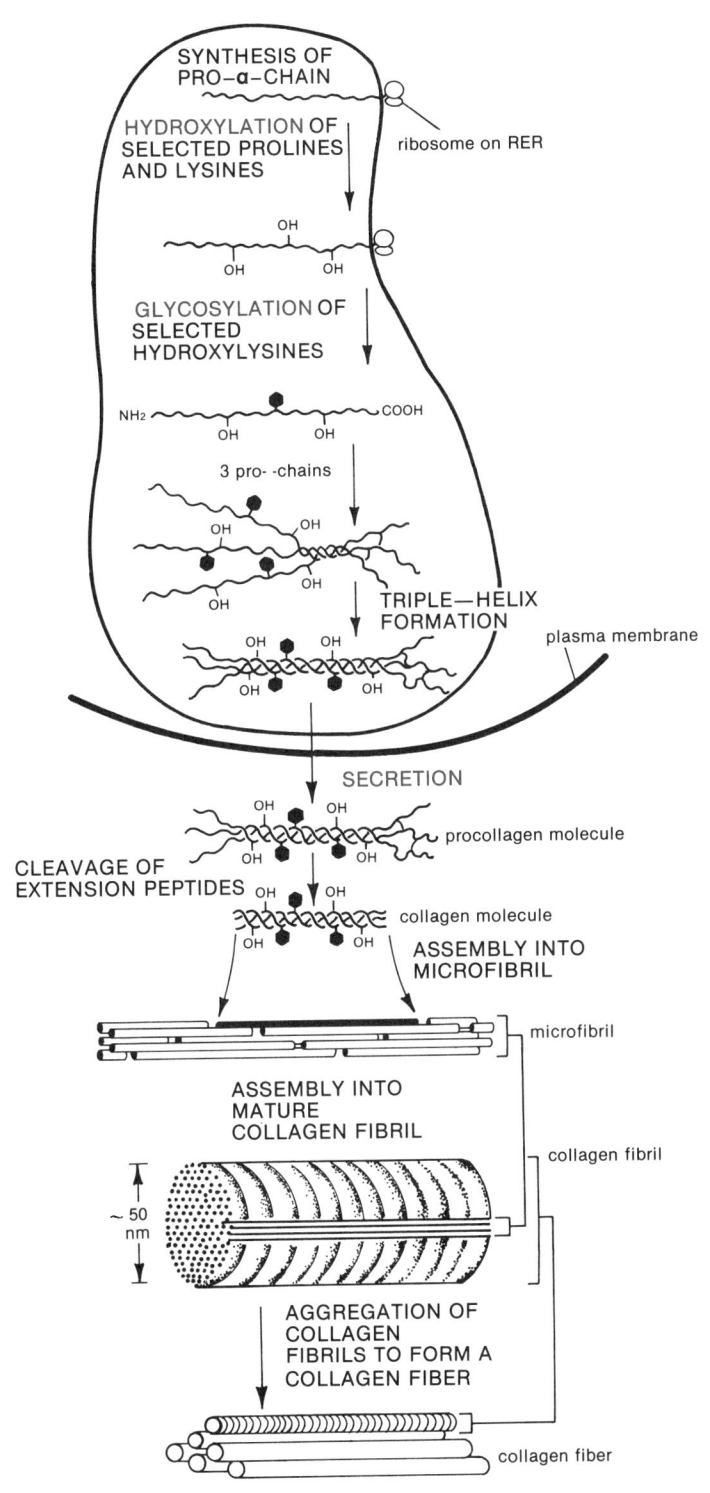

Posttranslational Modifications and Fibrillogenesis

Immediately after synthesis of the pro-α chains, as they are being extruded into the endoplasmic reticulum cisternal space, over 100 amino acids in each chain are chemically modified. **Proline hydroxylases** selectively hydroxylate certain prolines to make either 3- or 4-hydroxyproline. Another important enzyme, **lysine hydroxylase,** converts certain lysines to hydroxylysine. These enzymes have an absolute requirement for ferrous iron, molecular oxygen, and ascorbic acid (vitamin C). The molecular defect in **scurvy** is deficient hydroxylation of collagen as a result of ascorbic acid deficiency.

As the lysines are hydroxylated, enzymes add sugar residues to some of the hydroxyl groups. These sugar additions, called glycosylations, are catalyzed by the enzymes **glucosyltransferase** (adding glucose), **galactosyltransferase** (adding galactose), and **mannosyltransferase** (adding mannose). Following hydroxylation and glycosylation, the pro-α chains polymerize into triple-helical procollagen (Fig. 4.4). The time necessary for synthesis of a complete chain is 6.7 minutes.

Mature procollagen molecules pass from the endoplasmic reticulum to the Golgi apparatus for final processing and packaging prior to secretion. The Golgi releases secretory vesicles full of procollagen ready for further extracellular modification and assembly. A compound called monensin, currently used as a research tool, blocks the export of procollagen and causes it to accumulate in Golgi vesicles.

Secretion of procollagen into the extracellular space initiates another series of reactions collectively known as **fibrillogenesis,** the final result being the appearance of collagen fibrils. Immediately after secretion, the enzyme **N-proteinase** cleaves the N-propeptides, and another enzyme, **C-proteinase,** removes the C-propeptides. Tropocollagen then self-assembles into fibrils. This occurs near the cell surface and is undoubtedly under some type of cellular control, not a random polymerization.

Hodge and Petruska discovered that the tropocollagen molecules assemble with a specific axial register. Certain molecular sequences along the triple helix direct the tropocollagens to line up in a quarter-stagger arrangement. They join such that the C-terminal heads are staggered by exactly 67 nm. This precise staggering leaves a gap between the head and the tail of consecutive molecules, and the gaps are in register across the entire thickness of the fibril, creating the characteristic cross-striation bands that are 67 nm apart (Fig.4.5).

The fibril is a staggered array of tropocollagen, and individual tropocollagen molecules within a fibril are **cross linked** to maintain integrity of the fibril and increase tensile strength. The important enzyme in cross linking is **lysyl oxidase,** which catalyzes an oxidative deamination of certain lysine residues in each tropocollagen molecule. The resultant lysine aldehydes form

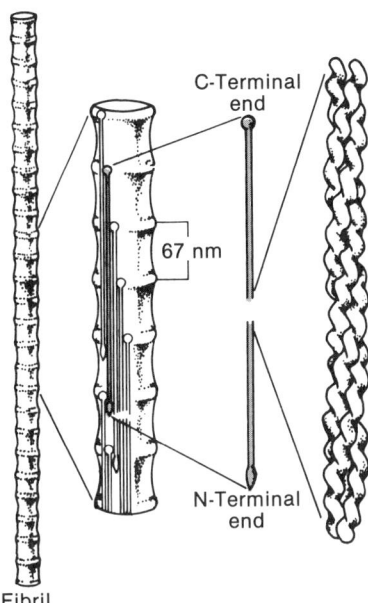

FIG. 4.5. The structure of a collagen fibril showing the staggered arrangement of tropocollagen molecules and the banding of a fibril. Only the interstitial collagens line up this way; type IV and type V collagens lack the necessary sequences. Adapted with permission from de Duve (1985).

Schiff bases ($R_1-C=N-R_2$) with adjacent molecules in the fibril. Cross links of this variety give collagen fibrils most of their tensile and torsional strength.

The importance of cross linking is apparent in a condition called **lathyrism,** a disease of animals caused by eating the seeds of sweet peas *(Lathyris odoratus)*. The seeds contain β-aminopropionitrile, which inhibits the cross-linking reaction, and the collagen of these animals is weak and very fragile, giving rise to scoliosis, severe limb deformities, and skin fragility.

Once they are cross linked, normal fibrils are quite stable and may reside in the extracellular matrix for many years, but eventually some of the fibrils are damaged and need removal and replacement. Removal of fibrils requires the participation of another set of special enzymes, the **collagenases.** Intact collagen is incredibly resistant to most enzymes, and only collagenases are capable of cleaving the helical region of native collagen. Tissue collagenase cuts tropocollagen at residue 750; the two fragments then unfold and are completely degraded by other proteolytic enzymes such as the **cathepsin** series (cathepsin D, B, G, N, L).

Disorders of Collagen Metabolism

In most diseases of connective tissue, collagen is damaged secondary to inflammation or the immune response, as in rheumatoid arthritis. In the heritable disorders of collagen, the tropocollagen molecule itself is abnormal, giving rise to a constellation of clinical signs and symptoms. These heritable disorders have been grouped into three major syndromes: Ehlers–Danlos syndrome, Marfan syndrome, and osteogenesis imperfecta.

People with **Ehlers–Danlos syndrome (EDS)** have fragile, stretchy skin and flexible ligaments giving hyperextensible joints (they are found as circus artists or limbo champions who can perform extraordinary contortions). Nine distinct varieties of EDS have been defined so far, but the disease is certainly heterogeneous since many patients remain with similar signs and symptoms that do not meet all the criteria for one of the known varieties.

In 1972, people with EDS VI were shown to have a deficiency of the enzyme lysyl hydroxylase, and their collagen was shown to lack hydroxylysine residues, forming inferior cross links. Since then, a defect in collagen structure or metabolism has been identified in the other varieties of EDS (Table 4.2). People with type IV EDS (associated with rupture of the aorta and of the large bowel) have defective type III collagen.

Since lysyl hydroxylase is a copper-containing enzyme, disorders of copper metabolism are similar to EDS. For instance, both Menkes kinky-hair syndrome and EDS IX are X-linked disorders, and they have abnormal serum copper and ceruloplasmin concentrations.

Marfan syndrome is an autosomal dominant disorder characterized by arachnodactyly, ligamentous laxity, excessive height, and ectopia lentis along with aortic and valvular disease. Abraham Lincoln may have had Marfan syndrome (a living direct descendant of Lincoln's uncle supposedly has the syndrome). In one variant of the Marfan syndrome, a defective pro-α 2(I) chain contains 20 extra amino acids, and other variants show a decreased synthesis rate of type I collagen.

Osteogenesis imperfecta (OI), also known as brittle bone disease, has

TABLE 4.2. *Molecular defects in heritable diseases of collagen*

Ehlers–Danlos syndrome	
Types I–III	Fibrillogenesis defects
Type IV	Decreased type III collagen
Type VI	Lysyl hydroxylase deficiency
Type VII	Persistence of N-propeptide
Type IX	Defective cross linking
Marfan syndrome	Abnormal pro-α 2(I) affecting structure of type I collagen
Osteogenesis imperfecta	
Type I	Probable deletion of d(1) gene
Type II	Defective secretion of α chains
Type III	Decreased pro-α 2(I) chains
Menkes syndrome	Cu metabolism abnormality causing defective cross linking

an incidence of one per 25,000 births. Osteogenesis imperfecta is divided into four general classes on the basis of inheritance and phenotype, but again, many patients do not accurately fit into the four types. Patients with OI show varying degrees of abnormality of tissues containing type I collagen. These abnormalities include brittle bones, lax ligaments, blue sclera, and defective tooth enamel. Prockop and other researchers have identified several distinct mutations that produce a change in the structure of type I collagen in patients with OI. Each mutation is at a different place, accounting for the heterogeneity of OI.

Type I OI has mildly brittle bones, blue sclera, and autosomal dominant inheritance. It is further subdivided by the presence or absence of dentinogenesis imperfecta. Type II OI is characterized by severe bone and extraskeletal involvement and is usually fatal at birth. Type III OI is the most variable phenotype, with moderately severe skeletal involvement, growth retardation, and autosomal recessive inheritance. Type IV OI is also variable but less severe, with normal sclera and dominant inheritance. Table 4.2 lists some molecular defects of type I collagen identified so far.

It may seem puzzling that such a broad range of phenotypic variations occur in genetic disease of type I collagen. However, considering the large number of posttranslational modifications and the complexity of the collagen molecule, the variability is not so surprising. Diseases of type I collagen can result from errors of biosynthesis associated with transcription, translation, or posttranslational modifications. A change in one part of the molecule might produce skin and eye disease, whereas a change in another part might produce bone manifestations. The number of possible genotypic changes still far outnumber the number of known phenotypic variations.

One of the more common disorders of collagen metabolism is **Dupuytren contracture,** in which the palmar fascia undergoes a progressive, irreversible contracture and nodule formation (the soles of the feet and the penis can also be involved). The contracture is not a simple shrinking of extraneous collagen fibers. A cell called the **myofibroblast** appears to be responsible for the disease. Myofibroblasts have characteristics similar to both fibroblasts and smooth muscle cells, making copious amounts of collagen but also possessing actin and myosin, which participate in contraction reactions. The combination of muscle-like contraction and connective tissue proliferation results in the characteristic clinical picture.

PROTEOGLYCANS

Early cytologists identified an "amorphous ground substance" surrounding cells of connective tissues. They noted the intense histologic staining with special cationic dyes such as safranin O and toluidine blue. Subsequent work from many laboratories has shown that the bulk of the amorphous ground substance is composed of certain macromolecules called **proteoglycans** (also known as mucoproteins), and the intense histologic staining is attributable to their polyanionic character, electrostatically bonding the cationic dyes.

The name proteoglycan reflects their chemical structure in that they are composed of 5 percent protein and 95 percent glycosaminoglycan. By weight, proteoglycans constitute a small part of the extracellular matrix, contributing 1 percent to the dry weight of bone, tendon, and ligament and 10 percent to that of articular cartilage. Their importance in the extracellular matrix far outweighs their minority weight status. Proteoglycans swell to fill the extracellular space and bind large quantities of the extracellular water, converting the matrix into a highly structured gel rather than an amorphous solution. They occupy vast amounts of space, and their large negative charge density electrostatically resists compression and contributes to the resilience and viscoelasticity of cartilage.

Glycosaminoglycans

The **glycosaminoglycans** (GAGs) are negatively charged linear polysaccharide chains composed of repeating disaccharide units. They are best characterized by composition of the disaccharide repeating unit and by position and amount of sulfation. One of the disaccharides is always an aminosugar, hence the name glycosamino; the name glycan is given to any large carbohydrate polymer. Negative charges come from the sugars possessing carboxylate groups. COO^-, and sulfate groups, SO^{4-}. The older literature often refers to glycosaminoglycans as acid mucopolysaccharides.

The glycosaminoglycans have enormous hydrodynamic volumes because of the water binding and the repulsion of close negative charges. None of the glycosaminoglycans occur as a single idealized disaccharide polymer. In fact, they show considerable heterogeneity in length as well as in degree of sulfation. Table 4.3 lists the six important glycosaminoglycans.

Hyaluronic acid is a copolymer of *N*-acetylglucosamine and glucuronic acid arranged as a single unbranched chain several thousand sugars long. It is the major GAG of synovial fluid and the vitreous humor, and, as will be discussed later, it is the central organizer of proteoglycan aggregates. Hyaluronic acid has a special function in embryogenesis and wound repair, causing matrix swelling and facilitating cell migration. Release of the enzyme hyaluronidase with resulting degradation of hyaluronic acid is associated with cessation of cellular migration.

Chondroitin-4- and **chondroitin-6-sulfate** are copolymers of *N*-acetylgalactosamine and glucuronic acid that differ only in the 4 or 6 position of sulfation. These are the most abundant GAGs in the body. Chondroitin-4-sulfate is found in embryonic tissue and immature cartilage, whereas chondroitin-6-sulfate is known to increase in cartilage with maturation and age.

STRUCTURAL COMPONENTS

TABLE 4.3. *The glycosaminoglycans*

GAG	Molecular weight	Repeating disaccharide	Tissue distribution
Hyaluronic acid	4,000 to 8 million	Glucuronic acid N-Acetylglucosamine	Synovium Cartilage Skin
Chondroitin sulfate	5,000 to 50,000	Glucuronic acid N-Acetylgalactosamine	Cartilage Skin Bone
Dermatan sulfate	15,000 to 40,000	Iduronic acid N-Acetylgalactosamine	Skin Vessels Heart
Heparan sulfate	5,000 to 12,000	Iduronic acid N-Acetylglucosamine	Cell surface Lungs Arteries
Heparin	6,000 to 25,000	Iduronic acid N-Acetylglucosamine	Mast cells Lungs Liver
Keratan sulfate	4,000 to 19,000	Galactose N-Acetylglucosamine	Cartilage disk

Chondroitin sulfate chains are shorter than those of hyaluronic acid, measuring between 20 and 100 repeating units in length. Chondroitin sulfate binds calcium, and endochondral calcification is accompanied by a decrease in the concentration of chondroitin sulfate in the longitudinal cartilaginous trabeculae.

Dermatan sulfate is a copolymer of N-acetylgalactosamine and iduronic acid (an epimer of glucuronic acid). This is the major GAG found in skin, tendon, heart valves, and arterial walls. It has the capacity to associate closely with collagen fibers.

Heparan sulfate is found in the lung, blood vessels, and on the external surface of many cells, but it is not found in skeletal tissue. Heparan sulfate on the surface of fibroblasts mediates cell–cell adhesion. **Heparin** is synthesized and stored in mast cells. Heparin is similar to heparan except that it has more sulfate and acetyl groups. Heparin works by inactivating plasma antithrombin, which is an α-globulin that inactivates thrombin. It also has other effects such as inhibiting the binding of C1q to immune complexes. The long-term use of heparin is known to be associated with osteopenia.

Keratan sulfate is a copolymer of N-acetylglucosamine and galactose, and it is the shortest of the GAGs. Keratan sulfate and chondroitin sulfate are the major glycosaminoglycans of cartilage proteoglycans.

Proteoglycan Subunits and Aggregates

The backbone of a **proteoglycan subunit** is a core protein that has a molecular weight from 200,000 to 350,000 (depending on the protein length).

Core protein contributes only 5 to 10 percent to the total mass of the proteoglycan subunit (most of the mass coming from the GAGs). Core proteins are also heterogeneous, but each has three distinct functional regions: (1) the hyaluronic-acid-binding region at the end, (2) a keratan-sulfate-rich region, and (3) a larger chondroitin-sulfate-rich region (Fig. 4.6).

The hyaluronic-acid-binding region is the globular end of the core protein that makes an elaborate fit with five disaccharides (10 sugars) of a hyaluronic acid molecule. The binding region will only associate with hyaluronic acid and none of the other glycosaminoglycans. This binding region is followed by the keratan-sulfate- and the chondroitin-sulfate-rich regions, where the component **GAG side chains** are attached to the core protein. About 60 percent of the keratan sulfates reside in the specific keratan sulfate region, with the remainder scattered throughout the molecule. Ninety percent of the chondroitin sulfate is found in the chondroitin sulfate region. The predominant proteoglycan from cartilage has about 130 keratan sulfate chains and about 100 chondroitin sulfate chains.

Each GAG side chain attaches to the core protein by a short four-sugar oligosaccharide linkage region consisting of xylose–galactose–galactose–glucuronic acid. Xylose covalently bonds to the free hydroxyl of a core protein serine, and glutamic acid bonds to the GAG side chains, fixing each GAG to the core protein. Electron micrographs show that a proteoglycan subunit is similar in appearance to a bottle brush, with an internal core protein and the GAG side chains extending, bristle-like, outward (Fig. 4.6).

Each proteoglycan subunit has a molecular weight in the range of one to four million. The variation of molecular weights arises from the extreme polydispersity in size and composition. There is no typical subunit, just aver-

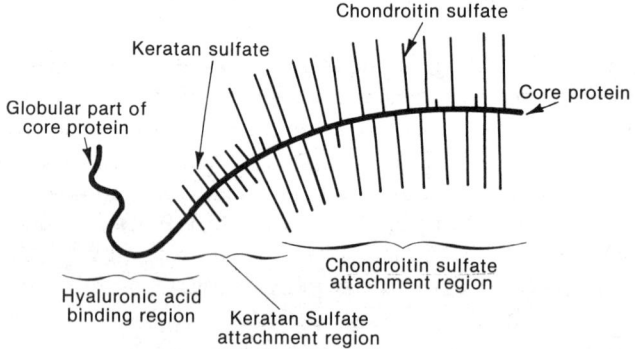

FIG. 4.6. Schematic diagram of a typical proteoglycan subunit (note the similarity to a bottle brush). The core protein is divided into three regions: the globular hyaluronic-acid-binding region, the keratin-sulfate-rich region, and the larger chondroitin-sulfate-rich region. The GAG chains attach to the core protein by covalent bonds with the free hydroxyl groups of serine residues. The repulsion of multiple negative charges on the GAG chains assure the maximum volume expansion of the subunit.

age subunits, which vary in core protein length, GAG length, and GAG proportions as well as degree of sulfation. One can understand why it was so difficult to work out the biochemistry of these molecules.

In the extracellular matrix, most proteoglycans exist as **proteoglycan aggregates.** The central organizing molecule of an aggregate is a long hyaluronic acid chain to which the subunits form noncovalent associations. These associations rely on the hyaluronic-acid-binding region of the proteoglycan subunit and are stabilized by **link proteins,** separate smaller proteins with molecular weights of 40,000 to 60,000. The size of a proteoglycan aggregate is determined by the length of the hyaluronic acid. In the usual situation, 100 to 500 proteoglycan subunits attach to a hyaluronate chain at intervals of about 300 angstroms (Fig. 4.7). The hyaluronic acid chain is anchored in turn to collagen fibrils scattered throughout the matrix.

Proteoglycan aggregates are extremely large biopolymers, often exceeding hundreds of millions of daltons (one dalton equals the mass of a hydrogen atom). The high negative charge density from the sulfates and carboxylates causes the aggregates to assume a shape that fills the maximum possible volume. Also, because of the hydrophilic structure and the charges, they retain and organize water into a shell of hydration trapped within and surrounding the aggregate, much like a sponge. Compression squeezes some of the fluid

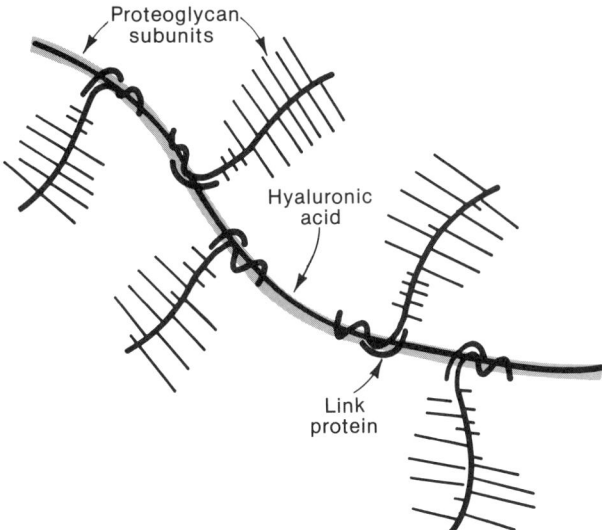

FIG. 4.7. Schematic diagram of a segment of proteoglycan aggregate. Numerous proteoglycan subunits are anchored to a central strand of hyaluronic acid. The subunits associate with the hyaluronic acid chain via the globular hyaluronic-acid-binding region, and each association is stabilized by link proteins. A typical proteoglycan aggregate from cartilage contains 100 to 150 subunits attached at intervals of 300 angstroms along the hyaluronic acid chain.

out of the proteoglycan domain, but the interacting negative charges strongly repel each other and resist further aggregate deformation. After removal of the deforming forces, the proteoglycan aggregate quickly expands and reabsorbs the water forced out by compression.

Comparing tropocollagen and proteoglycans, we can see how tropocollagen molecules best resist tension forces because they are packed tightly together and cross linked. Fibrils occupy a small volume and have great tensile strength. On the other hand, proteoglycan aggregates are greatly extended molecules and occupy the largest possible molecular domain. Their water content and electronegativity give them a high compressibility but a minimal tensile strength, ideal for resisting compressive forces such as those transmitted across joint surfaces.

Proteoglycans differ from tissue glycoproteins. Proteoglycans are 90 to 95 percent carbohydrate by weight, whereas glycoproteins are only 1 to 40 percent carbohydrate, which is present as short oligosaccharide chains rather than long repeating GAG chains.

Proteoglycans in articular cartilage have been shown to change with age, increasing in keratan sulfate and decreasing in chondroitin sulfate. The percentage of aggregation and the size of aggregates also decrease with aging. In the intervertebral disk, the proteoglycan concentration is highest in the nucleus pulposis and decreases toward the rim of the annulus. They are not as aggregated as proteoglycans found in articular cartilage. Bone contains smaller-molecular-weight proteoglycans not found in cartilage. In fact, proteoglycan aggregates inhibit calcification of cartilage and must be removed during endochondral ossification, as discussed in Chapter 3. Finally, certain forms of dwarfism are associated with inherited errors of proteoglycan biosynthesis (pseudoachondroplasia, diastrophic dwarfism, and brachymorphism).

To summarize, proteoglycans have the potential for tremendous heterogeneity. The typical cartilage proteoglycan subunit has about 100 chondroitin sulfates and 60 keratan sulfates attached to a serine-rich core protein of around 2,000 amino acids. The subunits bind to a hyaluronic acid chain and are stabilized by link proteins, forming a proteoglycan aggregate that occupies a large volume and organizes the surrounding water into an amorphous gel.

Synthesis, Degradation, and the Mucopolysaccharidoses

Synthesis of a proteoglycan involves over 30 separate enzymes performing numerous posttranslational modifications. Core proteins are synthesized on membrane-bound ribosomes and subsequently transferred to the smooth ER and to the Golgi. Here, monosaccharides are added sequentially, beginning at the oligosaccharide linkage region. The GAG chains are built up one sugar at a time from precursor UDP-sugars by specific enzymes called glycosyltransferases. The GAG chains are modified by sulfation and epimerization (changing the configuration around one carbon atom) prior to secretion of the assembled proteoglycan into the extracellular matrix.

Worn out or damaged proteoglycans in the matrix are degraded intracellularly. Protein fragments and GAG chains reenter the cell by endocytosis and are degraded sequentially by lysosomal enzymes. Complete degradation of proteoglycans requires many enzymes, and all the necessary enzymes have been found in lysosomes (including proteinases, specific glycosidases, and sulfatases). A deficiency of any one of the enzymes for GAG degradation produces a storage disorder known as a **mucopolysaccharidosis.**

Absence or deficiency of a degradation enzyme blocks any further degradation at that step. Incompletely degraded chains accumulate in the lysosomes and spill over into serum and into the urine. For instance, **Morquio syndrome** is caused by a block in keratan sulfate breakdown and can be diagnosed by detecting keratan sulfate in the urine of an individual with the characteristic phenotype. The enzymatic defect in **Hurler syndrome** is an absence of sulfoiduronate sulfatase, so sulfated GAG chains massively accumulate in the liver, spleen, and other organs. Table 4.4 lists the clinical features of some mucopolysaccharidoses.

ELASTIN, FIBRONECTIN, AND OTHER PROTEINS

Elastin is a highly cross-linked protein polymer that can be stretched to several times its length and then return to its original size and shape when the tension is released. Under the microscope, elastin has a wavy appearance because of the random coil orientation of the monomers. When it is stretched, the waviness disappears, and the monomers assume a more parallel orientation (Fig. 4.8). Large amounts of elastin are found in arterial walls and in certain ligaments such as the ligamentum nuchum and the ligamentum flavum. Elastin is also a major component of elastic cartilage in the ear and in the epiglottis.

The basic monomer of elastin is **tropoelastin,** a single polypeptide chain of about 800 amino acid residues with a molecular weight of 68,000. As in the case of collagen, glycine accounts for one-third of the amino acids in elastin, and proline is present in high amounts. However, the similarity ends there. Elastin contains very little hydroxyproline and no hydroxylysine. Furthermore, most of the amino acids are hydrophobic as compared to the preponderance of hydrophilic amino acids in collagen.

Tropoelastin is secreted into the extracellular space, where it is assembled and cross linked into elastin fibers. All of the cross links are derived from lysine, either by aldol cross links via Schiff bases or by the combination of four lysines to make **desmosine.** Desmosine is a heterocyclic nitrogen com-

TABLE 4.4. *The enzymatic and clinical features of the most common mucopolysaccharidoses*

Disease	Enzyme defect	Excessive accumulation	Clinical features
Hurler syndrome MPS I	Iduronidase	Dermatan and heparan sulfate	Misshapen, short bones Clouded cornea Coarse facies Mental retardation Hepatosplenomegaly
Hunter syndrome MPS II	Iduronate sulfatase	Dermatan and heparan sulfate	Broad bones Stiff joints Hepatosplenomegaly
Sanfilippo B MPS III B	N-Acetylglucosaminidase	Heparan sulfate	Mental retardation Short stature Hepatomegaly
Morquio syndrome MPS IV	N-Acetylgalactosamine 4-sulfatase	Keratan sulfate	Short stature Severe kyphosis Genu valgus Clouded cornea
Maroteaux–Lamy syndrome MPS VI	N-Acetylgalactosamine 6-sulfatase	Dermatan sulfate	Coarse facies Stiff joints Short stature No mental deterioration

pound that forms as a result of tetrafunctional cross links among lysine residues on four elastin monomers (Fig. 4.8).

Elastin is an extremely stable compound, resistant to physical stress, proteolytic digestion, and most acid and alkaline degradation reactions. It is degraded *in vivo* by the enzyme **elastase,** found in high concentrations in pancreatic juice and also in polymorphonuclear leukocytes. Elastase is a very aggressive enzyme, degrading not only elastin but also some collagen and many other tissue proteins. This is a potentially dangerous enzyme in the body, and the extracellular activity is closely controlled by a special α-1-proteinase inhibitor (previously known as α-1-antitrypsin). The absence of this inhibitor results in early, severe emphysema with skin and arterial changes from the unrestricted activity of elastase.

The **fibronectins** are a family of glycoproteins found in the extracellular matrix and also present in the plasma. Fibronectins (L. *fibra,* fiber, and *nectene,* connect) are disulfide-linked dimers of 450,000 daltons. They simultaneously bind to cell surface receptors and to other structural components, connecting and anchoring the cells and proteins in the extracellular matrix.

Shaped like a giant V, the fibronectin molecule has specific binding domains along the arms of the V. Binding domains include the cell surface binding domain, the collagen domain, the hyaluronic acid domain, and the glycosaminoglycan domain. These multiple binding domains connect colla-

STRUCTURAL COMPONENTS

gen to glycosaminoglycans, cells to collagen, etc., resulting in an organized, interconnected meshwork.

Fibroblast motility requires fibronectin because the cells must grip fibronectin-containing components as they move about the matrix. Fibronectin is also involved in wound healing, embryogenesis, and malignancy. Fibronectin-rich pathways guide the migration of many kinds of embryonic cells. Malignant cells have a greatly reduced fibronectin-binding ability, contributing to the loss of contact inhibition.

Chondrocytes use another glycoprotein called **chondronectin** for matrix adhesion. Chondronectin is smaller and less complex than fibronectin, with a molecular weight of 180,000. Osteocytes utilize **osteonectin,** and epithelial

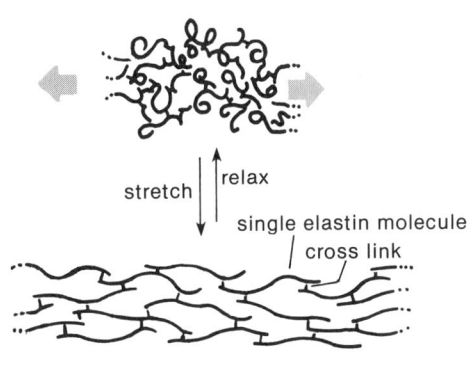

FIG. 4.8. A: Elastin monomers are joined together by extensive cross linking. The entire network can stretch and recoil like a rubber band. **B:** The structure of desmosine, which is formed from four lysine side chains on four elastin molecules.

cells attach to the matrix with yet another protein called **laminin**. This is a larger glycoprotein of 800,000 daltons, which specifically binds to type IV collagen.

MINERAL

Composition of Bone Mineral

Bone mineral has many chemical characteristics similar to the geologic mineral **hydroxyapatite** $[Ca_{10}(PO_4)_6(OH)_2]$. X-ray crystallography shows that bone mineral is a poorly crystalline analogue of hydroxyapatite and differs from the geologic mineral in the following ways: (1) bone mineral is calcium deficient and contains carbonate ions; (2) the atomic arrangement is imperfect, with vacant lattice slots randomly occupied by carbonate, magnesium, fluoride, or other ions; (3) the crystals are small rod-shaped structures, 100 to 400 angstroms in length, which are slightly more soluble than true hydroxyapatite.

The composition of bone mineral changes with maturation and age. Initial new bone resembles calcium hydrogen phosphate dihydrate, also known as the mineral **brushite** ($CaHPO_4 \cdot 2H_2O$). Acid phosphate groups (HPO_4^{2-}) in the new bone crystals disappear with maturation, and the calcium content increases.

Ions of magnesium, strontium, and lead are readily incorporated into the crystal lattice. The lead lines of children who have eaten lead-based paint are concentrations at the mineralization front, and the incorporation of radioactive strontium 90, produced in nuclear power reactors and radiation fallout, is a potential health hazard. A more beneficial situation results from the substitution of fluoride for some of the hydroxyl groups in the lattice. Fluoride decreases the solubility and increases the stability of the crystal, and this is the basis for fluoride treatment to prevent tooth decay and to treat osteoporosis.

Pyrophosphates, compounds with the structure $P-O-P$, inhibit crystal formation by changing the surface geometry of the growing crystal. The diphosphonates contain the structure $P-C-P$, and they also interfere with crystal growth. Ethane-1-hydroxy-1,1-diphosphonate (EHDP) and dichloromethylene diphosphonate (Cl_2MDP) have been used clinically with limited success in Paget's disease of the bone, heterotopic ossification, and hypercalcemia of malignancy.

Osteoid

Osteoid is the organic matrix of bone synthesized by osteoblasts prior to mineral deposition. Type 1 collagen accounts for 70 percent of osteoid (90 percent of the dry weight). The remainder consists of bone proteoglycans and noncollagenous proteins. The major noncollagenous proteins are osteonectin, bone morphogenetic protein, bone phosphoproteins, bone sialoprotein, and osteocalcin.

As discussed in the section on collagen, above, collagen molecules line up in a quarter stagger array, resulting in gaps (hole zones) in the fibrils. Early bone mineral occupies the hole zones, and as the mineral crystals grow, they align parallel to the long axis of the collagen fibril. **Osteonectin** binds to type 1 collagen, to mineral crystals, and to calcium salts. It is thought to be involved in mineral orientation and regulation of mineral growth. **Bone morphogenetic protein** is a glycoprotein that has osteoinductive properties; that is, it causes the recruitment and activation of bone-forming cells. **Phosphoproteins** and **osteocalcin** apparently regulate ossification by inhibiting mineral formation. Osteocalcin contains a unique amino acid, γ-carboxyglutamic acid, and the rate of excretion of γ-carboxyglutamic acid is a specific indicator of bone turnover. Osteocalcin synthesis has an absolute requirement for vitamin K. The exact role of the sialoproteins, bone proteoglycans, and other proteins is currently an area of active research.

Mineralization of Bone

Calcium and phosphate concentrations in serum and extracellular fluids are metastable with regard to bone mineral. This means that calcium and phosphorus will not spontaneously precipitate, but the addition of a small crystal nidus results in a rapid precipitation and mineral crystal proliferation. Mineralization and bone formation occur when the equilibrium shifts from metastable to unstable, and calcium salts begin to precipitate. This process can be considered in four stages: (1) matrix modification; (2) crystal nucleation; (3) crystal growth; (4) remodeling.

Matrix modification occurs in both mineralizing growth cartilage and osteoid, although the cells, collagen, and matrix proteins are different. The process appears to be, at least in part, under control of the chondrocytes and osteoblasts. Local calcium and phosphate concentrations increase, probably as a result of export of calcium from mitochondria and release of phosphorus from the breakdown of phosphoproteins and pyrophosphate. Matrix vesicles containing alkaline phosphatase and pyrophosphatase bud off the plasma membrane of chondrocytes or osteoblasts, and inhibitory macromolecules such as aggregated proteoglycans are removed from the matrix.

The next event is **crystal nucleation,** during which the first small mineral crystal appears. The exact mechanism is obscure, but crystal formation has been observed inside matrix vesicles and also in the hole zones of collagen fibrils. **Crystal growth** proceeds rapidly and is oriented by the orientation of collagen fibrils. **Remodeling** begins immediately by osteoclastic resorption coupled with new bone formation.

TABLE 4.5. *Types of biologic mineralization*

Ossification	Bone formation (the presence of osteones) by the endochondral or intramembranous mechanism
Heterotopic ossification	The presence of bone in a nonphysiological place such as muscle or bladder mucosa
Tumor calcification	Calcium deposits and bone formation in a matrix produced by malignant cells
Dystrophic calcification	Calcium deposits in damaged tissues that are perfused with extracellular fluids of normal calcium and phosphorus concentrations such as calcific bursitis and aortic valve calcification
Metastatic calcification	Calcium phosphate deposition caused by elevation of the serum Ca^{2+} or PO_4 concentrations, as in hyperparathyroidism or renal disease
Crystal deposition disease	The deposition of certain inorganic ions in soft tissues because of a metabolic abnormality such as chondrocalcinosis, hemochromatosis, or gout

The terms **mineralization, calcification,** and **ossification** are often confused. In mammals, mineralization and calcification are essentially identical terms describing the deposition of inorganic mineral in biological substrates; the terms may be used synonymously. On the other hand, ossification is a special term describing mineral deposition and cellular activity resulting in the formation of bone (Table 4.5).

GLOSSARY

α chain   A polypeptide chain of 1,055 amino acids, three of which are coiled into a tropocollagen molecule.
Basal lamina collagen   Type IV collagen found in basement membranes in association with proteins such as laminin.
Bone morphogenetic protein   A glycoprotein present in bone matrix that has the ability to recruit and activate bone-forming cells.
Brushite   A geologic mineral of the general composition $CaHPO_4 \cdot 2H_2O$ thought to resemble newly formed bone mineral.
Desmosine   A heterocyclic nitrogen compound found in elastin and derived from the side chains of four lysine residues on four tropoelastin monomers, serving to cross link the monomers.
Elastin   A highly cross-linked protein constructed of monomers of tropoelastin and capable of great elastic deformation.
Fibrillogenesis   The extracellular process of maturation and assembly of tropocollagen molecules into fibrils.
Fibronectin   A family of extracellular proteins that act to connect and anchor cells and various extracellular macromolecules.
Glycosaminoglycans   Linear polysaccharide chains composed of repeat-

ing disaccharide units, many of which contain sulfate groups or free carboxylate groups, previously called mucopolysaccharides.

Hyaluronic acid The largest glycosaminoglycan, HA is a copolymer of N-acetylglucosamine and glucuronic acid and is the central organizing backbone of proteoglycan aggregates.

Hydroxyapatite A geologic mineral of the general composition $Ca_{10}(PO)_6(OH)_2$ that has crystallographic similarities to bone mineral.

Interstitial collagens Those collagens that are found as an interstitial component of connective tissues, namely, type I in bone, tendon, ligament, and skin, type II in cartilage, and type III in arteries, muscle, and organs, also called fibrillar collagens.

Matrix vesicles Extracellular membrane-bound structures derived from osteoblast and chondrocyte plasma membranes containing alkaline phosphatase, inorganic pyrophosphatase, and ATPase, serving as an initial locus of mineralization in growth cartilage and other tissues.

Mineralization The general process of ion deposition in biological tissues, in mammals; used synonomously with calcification.

Mucopolysaccharidoses Storage diseases resulting from the absence or deficiency of a degradative enzyme in the glycosaminoglycan pathway.

Osteocalcin A noncollagenous matrix protein containing γ-aminoglutamic acid and acting as an inhibitor of mineral crystal growth.

Osteogenesis The overall process of bone formation involving differentiation of bone cells from mesenchymal precursors, osteoid formation, and mineralization of the matrix followed by remodeling to produce bone.

Osteoid The organic matrix of bone, synthesized and secreted by osteoblasts, composed of type I collagen, various noncollagen proteins, and bone proteoglycans.

Osteonectin A noncollagenous osteoid protein, analogous to fibronectin, that binds type I collagen and mineral crystals.

Procollagen The precursor protein of tropocollagen, containing bulky N-terminal and C-terminal extension peptides.

Proteoglycans Matrix macromolecules that have the general shape of a bottle brush, heterogeneous in structure, consisting of a core protein and attached GAG chains. They can occur in isolation as subunits or as large aggregates.

Tropocollagen A basic collagen molecule consisting of a triple helix of three α chains, with molecular dimensions of 300 $\times$ 1.5 nanometers and having a molecular weight of 285,000.

BIBLIOGRAPHY

Albers, B., Bray, D., Lewis, J., Raff, M., Roberts, K., Watson, J. (1983): Molecular Biology of the Cell. Garland, New York.

Anderson, H. C. (1980): Calcification processes. *Pathol. Annu.*, 15:45–75.

Boskey, A. L., and Posner, A. S. (1984): Bone structure, composition, and mineralization. *Orthop. Clin. North Am.,* 15:597–612.
Buckwalter, J. A. (1983): Proteoglycan structure in calcifying cartilage. *Clin. Orthop.,* 172:207–232.
de Duve, C. (1985): *A Guided Tour of the Living Cell.* W. H. Freeman, San Francisco.
Eyre, D. R. (1981): Concepts in collagen biochemistry: Evidence that collagenopathies underlie osteogenesis imperfecta. *Clin. Orthop.,* 159:97–107.
Glimcher, M. J. (1984): Recent studies of the mineral phase in bone and its possible linkage to the organic matrix by protein-bound phosphate bonds. *Phil. Trans. R. Soc. Lond.* [*Biol.*], 304:479–508.
Glorieux, F. H. (1982): Mineral. In: *The Musculoskeletal System. Embryology, Biochemistry, and Physiology,* edited by R. L. Cruess, pp. 97–106. Churchill Livingstone, New York.
Hassell, J. R., Kimura, J. H., and Hascall, V. C. (1986): Proteoglycan core protein families. *Annu. Rev. Biochem.,* 55:539–567.
Mosher, D. F. (1984): Physiology of fibronectin. *Annu. Rev. Med.,* 35:561–575.
Nimni, M. E. (1983): Collagen: Structure, function, and metabolism in normal and fibrotic tissues. *Semin. Arthritis Rheum.,* 13:1–8.
Posner, A. S. (1985): The mineral of bone. *Clin. Orthop.,* 200:87–99.
Prockop, D. J. (1986): Genetic defects of collagen. *Hosp. Pract.,* 21:125–140.
Prockop, D. J., and Kivirikko, K. I. (1984): Heritable diseases of collagen. *N. Engl. J. Med.,* 311:376–386.
Prockop, D. J., Kivirikko, K. I., Tuderman, L., and Guzman, N. D. (1979): The biosynthesis of collagen and its disorders. *N. Engl. J. Med.,* 301:13–23, 77–85.
Sandberg, L. R., Soskel, N. T., and Leslie, J. G. (1981): Elastin structure, biosynthesis, and relation to disease states. *N. Engl. J. Med.,* 304:566–579.
Seyer, J. M., and Kang, A. H. (1985): Structural proteins: Collagen, elastin, and fibronectin. In: *Textbook of Rheumatology,* edited by W. N. Kelley, E. D. Harris, S. Ruddy, and C. B. Sledge, pp. 211–237. W. B. Saunders, Philadelphia.
van der Rest, M. (1982): Collagen structure and biosynthesis. In: *The Musculoskeletal System. Embryology Biochemistry and Physiology,* edited by R. L. Cruess, pp. 59–79. Churchill Livingstone, New York.
Wuthier, R. E. (1952): A review of the primary mechanism of endochondral calcification with special emphasis on the role of cells, mitochondria, and matrix vesicles. *Clin. Orthop.,* 169:219–242.

5
Bone Morphology and Biology

BONE AS MATERIAL, A TISSUE, AN ORGAN

Bone can be studied in three ways: as a material, as a tissue, and as an organ. Bone as a **material** is a **compound composite** that is very resistant to compression—similar to reinforced concrete, with the collagen fibrils representing the steel rods and the mineral matrix as the cement. Studied as a material, bone has remarkable properties. It is strong yet light, equal in bending strength to oak and in tensile strength to cast iron but one-third as heavy. Three factors contribute significantly to the material properties of bone: (1) laminar construction of the cortex and trabeculae, (2) tubular design of long bones with radial distribution of the mass, and (3) internal reinforcement by a trabecular mesh.

Bone as a **tissue** consists of living cells embedded in a highly vascular, mineralized, osteoid matrix. Osteoblasts lay down osteoid seams with an average width of 12 micrometers. Bone is 92% solid and 8% water. Sixty-five percent of bone dry weight is mineral, and 35% is matrix. Collagen is the major organic matrix component at 95 percent. Osteocytes, osteoblasts, and osteoclasts make up about 3% of bone volume. Table 5.1 lists the general composition of bone tissue.

Bone as an **organ** contains many different tissues such as cartilage, nerve, fibrous tissue, marrow, adipose tissue, and vascular tissue. As an organ, bone changes with age. Young bones are porous, flexible, and contain red hematopoietic marrow. Old bones are more rigid and have a fibrofatty marrow.

Genetics determines the overall shape of a bone. This is apparent from the shape of embryonic anlagen that have a remote but obvious resemblance to their adult counterparts. During somatic growth and remodeling, the bone is influenced by both physical and biological stresses (weight bearing, muscle pull, nutrition, etc.), and the adult bone is a time-averaged solution to all these stresses.

Bones serve a variety of different functions. (1) They act as internal sup-

TABLE 5.1. *The general composition of bone*

Solids	92%
Water	8%
Solid composition	
Mineral phase	65%
Organic phase	35%
Mineral phase composition	
Calcium	60%
Phosphorus, Mg, Na, other ions	40%
Organic phase composition	
Collagen	95%
Cells	3%
Lipids, glycosaminoglycans, noncollagen proteins, etc.	2%

ports for the trunk and extremities and as a scaffolding for the attachment of muscles and ligaments. (2) Bones protectively cover the brain, spinal cord, and also the thoracic organs. (3) They provide a home for hematopoietic tissue. (4) Bones function in mineral ion homeostasis and as an ion reservoir containing 99% of the body calcium, 85% of the phosphorus, 66% of the magnesium, and 60% of the sodium. Bones are living, changing organs of support, locomotion, protection, and metabolism, constantly responding and remodeling to internal and external stimuli.

CLASSIFICATION OF BONE TISSUE

Bone may be classified on the basis of shape, macroscopic appearance, developmental biology, or microscopic organization. Each classification has certain advantages and disadvantages, but no one scheme completely satisfies all scientific and clinical needs. Thus, it is useful to review all four classifications (Table 5.2).

External geometry is the basis for the **shape classification.** Bones are grouped as long or short, flat or tubular, regular or irregular. The simplicity of this scheme limits its usefulness except for elementary descriptions.

TABLE 5.2. *A classification of bone*

Shape	Long, irregular
	Short, flat
Macroscopic	Cortical (compact)
	Cancellous (trabecular)
	Fine cancellous (embryonic)
Developmental	Membranous
	Endochondral
Microscopic	Woven
	Lamellar (Haversian and non-Haversian)

Visual inspection is the basis for a **macroscopic classification** of bones into three types: (1) **cortical,** (2) **cancellous,** and (3) **fine cancellous.** Cortical (or compact) bone comprises the diaphysis of long bones and the cortex of all others. Most of the skeletal mass (80% by weight) is cortical bone, but most of the skeletal volume is cancellous bone. Cancellous (or trabecular) bone occurs in the metaphysis of long bones and in the interior of flat bones such as the scapula or the innominate. Cancellous bone comprises 20 percent of the mass but has a very large surface area relative to both mass and volume. Fine cancellous bone, found only in the embryo, makes up both the diaphysis and metaphysis of newly ossified embryonic bone.

Origin is the basis for a **developmental classification** of bones, and there are two types: (1) **intramembranous** and (2) **endochondral.** Intramembranous bones (skull and clavicle) are formed by osteoblasts that differentiate directly from a membranous periosteum. Endochondral bones, by far the majority, initially appear as cartilage models that are converted to bone.

Matrix and cellular arrangement form the basis for a **microscopic classification** of bone into two types: (1) **woven bone** and (2) **lamellar bone.** Woven bone is the first bone made at the growth plate, in fracture callus, and in certain pathological conditions such as Paget's disease. Cells and collagen assume a random arrangement in woven bone. The matrix is cellular and stains unevenly and patchily with basophilic dyes. Lamellar bone replaces woven bone in the process of remodeling (Fig. 5.1).

Lamellar bone makes up the cortex and trabecula of mature bone and is

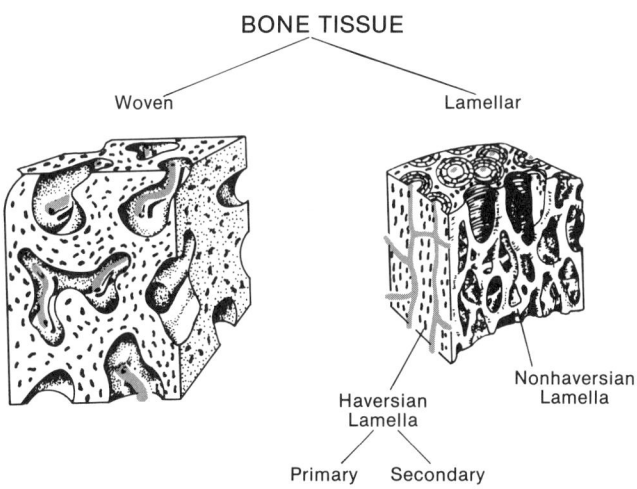

FIG. 5.1. Bone tissue is either woven or lamellar, and lamellar tissue is further subdivided into Haversian lamellae (in cortical bone) or non-Haversian lamellae (in cancellous bone). Primary Haversian lamellae are replaced by secondary Haversian lamellae during remodeling.

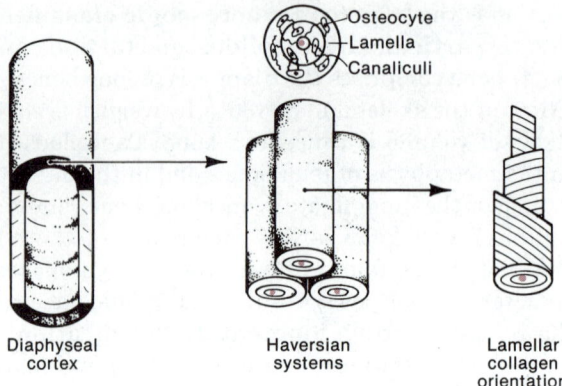

FIG. 5.2. The diaphyseal cortex of a long bone is composed of Haversian systems built around a central canal. The osteocytes lie within lacunae and communicate with the central canal via cytoplasmic processes running in canaliculi. Collagen fibrils have a different orientation in each lamella, adding to the composite strength.

subdivided into Haversian lamellar bone and non-Haversian lamellar bone. All lamellar bone has an orderly cellular distribution and regularly arranged collagen fibrils laid down within osteoid seams. In each successive osteoid layer, the collagen fibrils differ in the angle of fibril orientation, similar to plywood in appearance and function (Fig. 5.2). Lamellae of Haversian bone are circumferentially arranged around a central vascular canal, and individual osteocytes communicate with the vessels in the central canal by cellular processes running through an elaborate system of canaliculi.

Cancellous or trabecular bone has few if any Haversian systems. The construction consists of three or four lamellae sandwiched together to form a three-dimensional osseous lattice arranged along lines of force. Since the trabecular width is small, canaliculi communicate directly with capillaries in the medullary canal, so blood vessels do not course through trabeculae; they run along the side.

BONE SURFACES AND MEMBRANES

Each long bone has three surfaces enclosed within an envelope of two membranes. The three surfaces are the periosteal surface, the endosteal surface, and the Haversian surface. The two membranes are the periosteum on the outside and the endosteum on the inside.

The **periosteum** is a two-layered membrane. The outer **fibrous layer** consists of dense white fibrous tissue and elongated, highly differentiated fibroblasts; the inner **cambium layer** is the germinal layer and contains reticulin and elastic fibers, fibroblasts, plus undifferentiated mesenchymal cells. The periosteum is thick and elastic in infants and children and overlays a layer of plump osteoblasts. In adults, it is thin and contains only occasional undifferentiated cells. Sharpey's fibers anchor the periosteum to bone.

The **endosteum** lines the inner or **endosteal surface** of a long bone and insulates bone from the marrow components. The endosteum is a highly vascular mat of delicate endothelial cells, capillaries, and reticular fibers.

A third surface, the **Haversian surface,** consists of lacunar and canalicular walls lined by osteocytes and osteocytic cellular processes. Both the endosteal surface and the Haversian surface are active in mineral metabolism.

This system of membranes isolates **bone extracellular fluid** from plasma. In other words, the skeleton is not in immediate electrolytic equilibrium with plasma or general extracellular fluid. In fact, the concentrations of electrolytes and proteins are quite different. For instance, potassium concentration is five times greater in bone extracellular fluid than in plasma, but calcium, magnesium, and sodium concentrations are lower. Albumen and other plasma proteins are excluded from bone fluid.

BONE STRUCTURE

Bones are a mixture of cortical and cancellous tissue. Cortical and cancellous tissue have identical composition but different morphology, as summarized in Table 5.3. In a long bone, cortical tissue of the diaphysis has the texture of ivory and is arranged as a hollow cylinder to best resist bending forces. The metaphysis flares to increase volume and surface area, thus decreasing stresses at contact points (joints). Cancellous tissue within the metaphysis is a meshwork of interlocking longitudinal and perpendicular plates arranged along internal lines of force. Cancellous plates support a thin layer of subchondral cortical bone and distribute weight-bearing and joint reaction forces into the bulk of bone tissue.

TABLE 5.3. *Cortical versus cancellous bone*

	Cortical	Cancellous
Skeletal mass	80%	20%
Mechanics	Rigid	Flexible
Cells	Osteocytes in lacunae	Osteocytes in lacunae
Osteocyte arrangement	Radial into osteones	Linear along trabeculae
Matrix	Mineralized osteoid	Mineralized osteoid
Blood supply	Penetrating vessels	Surface vessels
Surface area-to-volume ratio	Low	High
Mineral metabolism	Minor role	Major role

Both cortical and cancellous tissue are lamellar bone, but cortical tissue is Haversian lamellar bone, whereas cancellous tissue is non-Haversian lamellar bone. Haversian lamellae make up the functional units of cortical bone, the **Haversian system,** or the **osteone.** Each osteone is a series of concentric lamellae constructed around a central Haversian canal containing small arterioles, venules, and nerves. The long axis of the osteone is parallel to the long axis of the bone. Haversian canals communicate with the medullary cavity by transverse or oblique **Volkmann's canals.**

Within the circular lamellae, osteocytes live in the seclusion of their own separate lacunae, sending multiple cytoplasmic extensions running through canaliculi. The canaliculi are the lifelines of osteocytes, providing communication channels to other osteocytes and nutrition conduits to the central Haversian canal for metabolic exchange, ion metabolism, and hormone delivery.

Osteones vary in size and shape, but on the average each cylindrical osteone contains six concentric lamellae (range three to 15), and each lamella is 5 to 7 micrometers thick. The theoretical maximum diameter of an osteone is about 150 micrometers, a distance set by the limits of nutrient diffusion from the central canal.

Osteones also vary in degree of mineralization. Older osteones contain more mineral salts and show greater microradiographic density than young osteones, and the lamellae closest to the central canal contain more mineral than the peripheral lamellae.

A basophilic **cement line** delineates the peripheral limits of an osteone and bonds it to its neighbors. The cement line is 1 micrometer thick and has a different composition from the surrounding matrix, being high in glycosaminoglycan and noncollagen proteins. Both primary and secondary osteones have cement lines.

A **primary osteone (primary Haversian)** is the first Haversian system to be formed on calcified cartilage bars or in woven bone. Secondary osteones replace primary osteones during the remodeling process. **Secondary osteones (secondary Haversian systems)** develop from resorption tunnels made by cutting cones during the process of remodeling.

Non-Haversian lamellae make up trabeculae of the metaphysis and medullary canal plus the area between osteones, including the outer and inner surfaces of bone. Lamellae encircling the outside and inside of a bone are called the outer and inner circumferential lamellae. Osteones between the circumferential lamellae are interstitial lamellae.

BLOOD SUPPLY

Bone is a highly vascular tissue, receiving approximately 10 percent of the cardiac output. Blood supply varies according to the shape and size of a bone, but in general, blood enters a typical long bone through four distinct systems: the nutrient artery, the metaphyseal arteries, the epiphyseal arteries, and the periosteal arteries. In an adult, the periosteal arteries are atrophic, but abundant anastomoses occur between the other systems. Blood exits through nutrient, metaphyseal, epiphyseal, and periosteal veins. Immature bone has discrete metaphyseal and epiphyseal circulations separated by the growth plate (see Chapter 3).

Under normal circumstances, arterial flow through cortical bone is centrifugal, from the endosteal surface to the periosteal surface, and venous flow is opposite. A single nutrient artery penetrates the cortical shaft obliquely, pointing away from the dominant growing end of the bone. The nutrient artery is a vestige of the embryonic artery that pierced the cartilage anlagen to initiate the ossification process. Multiple metaphyseal and epiphyseal arteries penetrate the surface through numerous smaller foramina.

Once inside the medullary cavity, the nutrient artery divides into ascending and descending branches that continue to divide repeatedly and diverge. Terminal branches anastomose with metaphyseal and epiphyseal capillaries and eventually terminate in small loops beneath the subchondral bone. Radially oriented branches of the nutrient artery supply the cortex. These branches enter Volkmann's canals, which run perpendicular to the long bone axis, feeding the vessels in Haversian canals. Arteries and veins that are inside bone consist of a single layer of perivascular connective tissue. They do not need the multiple layers of adventitial tissue necessary for protection of extraosseous vessels, and they do not have a muscularis layer. Capillaries within Haversian canals have numerous fenestrations to facilitate irrigation of the canaliculi.

At the external bone surface, penetrating cortical vessels connect with the remnants of the periosteal plexus. The periosteal plexus in turn anastomoses with capillaries in overlying muscular cuffs. The periosteal system mainly provides a reserve supply. Blood flow through these vessels is normally minimal but increases and becomes centripetal after a fracture or following marrow injury (e.g., medullary reaming).

Small venous channels exit bone through all surfaces not covered by cartilage, but most blood returns from the cortex to enter the marrow sinusoids. The medullary venous system has six to eight times the capacity of the arterial system and is the point of circulatory entry for new blood cells from the hematopoietic system.

Lymphatics are found in the periosteum, but little is known about intraosseous lymphatics. Sensory and autonomic nerve fibers penetrate bone along with the vessels. The periosteum and periarticular tissue also have an extensive sensory innervation.

BASIC SKELETAL PROCESSES

Bones engage in five basic physiological processes during life, as pointed out by Harold Frost: (1) **growth,** (2) **modeling,** (3) **remodeling,** (4) **repair,** and (5) **blood–bone ion exchange.** Growth and modeling concern the enlargement of mass and volume as well as the change in shape. Bones accumulate mass and grow in width by periosteal apposition while the medullary canal expands by endosteal resorption. Length increases by endochondral ossification. Modeling is the change in architecture (both internal and external) by osteoblastic and osteoclastic activities of funnelization, cylinderization, and hemispherization (see Chapter 3). Growth factors, hormones, nutrition, genetics, and physical stresses all influence growth and modeling. After skeletal maturity, the rates of growth and modeling are negligible.

Remodeling

Remodeling is a process of bone removal and new bone replacement occurring at the periosteal surface, the endosteal surface, and within cortical and cancellous bone. It is a continual process of replacement in response to physical and metabolic demands. Remodeling is the mechanism by which function dictates structure of an adult bone **(Wolff's law)**. It was in 1868 that Julius Wolff proposed that every change in the form or function of a bone is followed by certain definite changes in its internal architecture and its external conformation.

The physiological concepts describing bone remodeling were proposed by Frost in 1966. Remodeling is a quietly active process, taking place only at small focal areas of bone at any one time but eventually involving the entire bone. In young adults, remodeling accounts for the turnover of about 10 percent of the skeleton per year. In both cortical and cancellous remodeling, osteoclastic bone resorption is coupled sequentially to osteoblastic bone formation; bone resorption is a prerequisite for new bone formation. Packets of cells called **basic multicellular units (BMUs)** carry out the coupled resorption–formation.

Remodeling by BMUs follows a sequence of (1) activation, (2) resorption, and (3) formation. The cycle begins with the activation of osteoclasts, forming a **cutting cone** of the BMU. Parathyroid hormone is the most potent osteoclast activator, but physical forces or simply normal turnover can also activate osteoclasts. The cutting cone excavates a tunnel roughly 400 micrometers wide and 2 to 10 millimeters long, completely removing some osteones and partially cutting across others. Bone formation occurs in the wake of osteoclastic resorption by cells of the **filling cone** (Fig. 5.3).

The filling cone is a collection of vascular tissue, undifferentiated cells, and osteoblasts trailing the cutting cone. The filling cone applies a cement line to the walls of the tunnel, demarcating the edge of resorption, and then osteoblasts refill the tunnel with successive layers of osteoid that begin to mineralize 8 to 10 days later. Vascular tissue remains centrally located as lamellae fill the tunnel.

Fluorescent markers such as tetracycline or radioactive precursors (^{32}P,

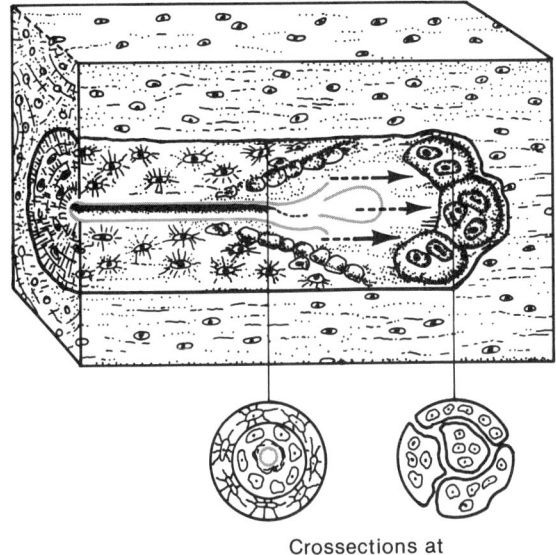

FIG. 5.3. Schematic diagram of a basic multicellular unit (BMU). The cross-sectional histology varies depending on which part is visualized. Osteoclasts on the right form a cutting cone that resorbs bone, followed by capillary buds and undifferentiated cells, the filling cone. Osteoblasts lay down osteoid and mature into osteocytes arranged as lamellae around the Haversian canal. Adapted from Perren and Rahn (1978).

^{45}Ca, ^{3}H-proline) permit an estimation of rates of new bone formation and BMU activity. Resorption by the cutting cone takes an average of 20 days, and refilling the tunnel with new bone takes another 80 days. Complete mineralization can take up to a year. Rates commonly vary depending on the age of the individual, the metabolic status, and local biophysical factors.

The principles of remodeling described for cortical bone also occur in cancellous bone, except the events take place on the surface of the trabeculae. Osteoclasts begin resorption in a cavity called the Howship's lacuna. The lacuna enlarges into a trench along the surface of the trabeculae. The cement line is applied, and osteoblasts fill in the trench.

Under homeostatic conditions, the rate of osteoclastic resorption equals the rate of new bone formation; bone tissue is renewed, but bone mass remains constant. In pathological states, the rates of resorption and formation dissociate, and the tissue balance changes. For instance, in senile osteoporosis, the rate of bone resorption is greater than the rate of new bone formation. Thus, the skeleton loses mass. In Paget's disease, the rate of bone turnover is greatly increased but without orientation. The ordered pattern of Haversian systems is replaced by a mosaic pattern of randomly arranged, irregular osteones.

Bone Growth Factors

Growth factors are organic modulators of bone growth that are not nutrients. It is convenient to divide growth factors into those that enhance bone resorption and those that enhance bone formation (Table 5.4).

The best-known factor that enhances bone resorption is parathyroid hormone, which directly inhibits osteoblasts and stimulates osteoclasts. Other activators of resorption include epidermal growth factor (EGF), fibroblast growth factor (FGF), and transforming growth factor (TGF). Both EGF and FGF stimulate bone resorption through a prostaglandin-dependent mechanism. Prostaglandins of the E series can directly activate osteoclasts. Transforming growth factor, found in both neoplastic and normal cells, has a similar mechanism of action and may be responsible for tumor-induced bone resorption. Another protein, called osteoclast-activating factor, is found in leukocytes and produces the endosteal bone loss in hematologic neoplasms.

Growth factors that enhance bone formation come from three sources: humoral factors, bone-derived factors, and tumor-derived factors. The humoral factors include insulin, growth hormone, and somatomedin C (also called IGF-1, which stands for insulin-like growth factor 1). Insulin switches on many biosynthetic enzymes and directly increases collagen synthesis in osteoblasts. Growth hormone works indirectly through somatomedin C (IGF-1). Somatomedin C is a small peptide (7,650 daltons) having "insulin-like" properties of increasing DNA and collagen synthesis and stimulating glucose oxidation, amino acid transport, and protein synthesis (Fig. 5.4).

The bone-derived growth factors characterized so far include bone morphogenetic protein (BMP), skeletal growth factor (SGF), and bone-derived growth factors (BDGFs). Cancellous and woven bone contain appreciable concentrations of these factors. Bone morphogenetic protein has a molecular

TABLE 5.4. *Growth factors and hormones affecting bone*

Enhancing resorption	Enhancing formation
Humoral factors	Humoral factors
Parathyroid hormone	Insulin
Glucocorticoids	Growth hormone
Local factors	Somatomedin C (IGF-1)
Epidermal growth factor	Thyroid hormone
Fibroblast growth factor	Bone-derived factors
Transforming growth factor	Bone morphogenetic factor
Osteoclast-activating factor	Skeletal growth factor
Prostaglandin E	Bone-derived growth factor
	Tumor-derived factors
	Prostatic carcinoma factor
	Breast carcinoma factor

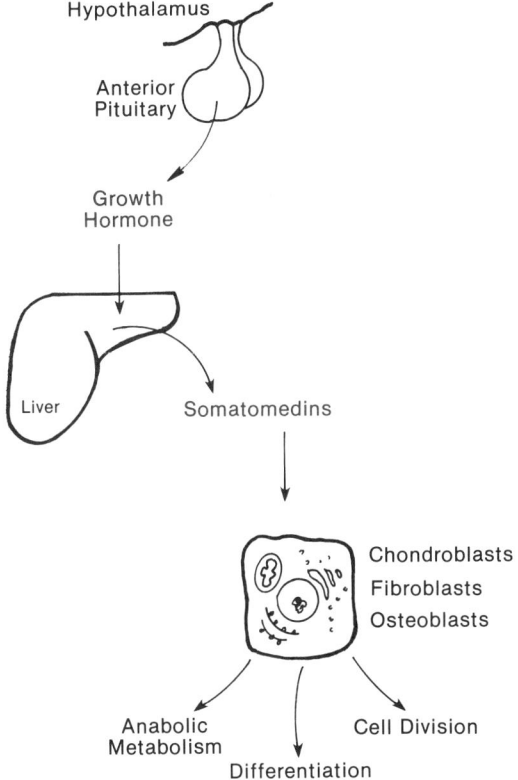

FIG. 5.4. Neuronal and chemical stimuli cause release of growth hormone-releasing factor into the hypophyseal portal circulation, causing the anterior pituitary to release growth hormone. At the liver, growth hormone stimulates the synthesis and release of somatomedins. Somatomedin C stimulates metabolism, differentiation, and division in chondroblasts, fibroblasts, and osteoblasts, resulting in tissue growth.

weight of 17,500 and stimulates new bone formation by inducing differentiation of perivascular mesenchymal cells into cartilage and woven bone. Skeletal growth factor is a larger protein of 83,000 molecular weight that stimulates cell division and collagen synthesis. The BDGFs also influence metabolic reactions such as DNA synthesis, hydroxyproline synthesis, and sulfate uptake. These agents differ in that BMP causes an irreversible differentiation of cells, but SGF and BDGFs produce a reversible mitotic stimulation comparable to the effects of somatomedin.

Metastatic prostate and breast cancer cells often stimulate new bone formation as well as bone destruction. Extracts of prostatic cancer cells contain a 67,000-molecular-weight protein that stimulates DNA synthesis and alkaline phosphatase production in osteoblasts. Breast cancer cells contain a similar protein. These neoplastic cells also contain transforming growth factors that activate bone resorption.

Hormones

Parathyroid hormone, insulin, growth hormone, and vitamin D are the most familiar hormones influencing bone growth as well as bone metabolism. Hormones work at the molecular level by influencing the synthesis of collagen, osteocalcin, osteonectin, or other phosphoproteins. At the cellular level, hormones affect matrix deposition and degradation, and at the tissue level, hormones influence the bone resorption–formation coupling ratio.

Lack of pituitary growth hormone causes dwarfism, whereas an excess of hormone produces gigantism (acromegaly after skeletal maturity). Excess glucocorticoids inhibit bone formation and decrease bone mass (cortisol-induced osteoporosis). Intestinal transport of calcium is decreased, and renal calcium losses are increased. Thyroid hormone is essential for normal growth and bone turnover. Decreased thyroid function retards endochondral ossification and delays closure of epiphyseal plates. Hypothyroid dwarfs have delayed appearance of secondary centers, and when the centers finally appear, they are fragmented or stippled (epiphyseal dysgenesis). Sex hormones influence growth and maturation and have profound protective effects against senile osteoporosis, again by unknown mechanisms.

BIOELECTRICITY

Bone is structurally dynamic, changing its internal and external shape by the mechanisms of remodeling outlined in the previous section. Both remodeling and skeletal repair are associated with regional electrical potential changes. In 1953, Fukada and Yasuda demonstrated that bone is **piezoelectric;** i.e., when bone is deformed, it generates an electrical potential. This potential is strain related and stress induced, and the magnitude of the potential is directly proportional to the amount of stress. An electrical potential difference between parts of a bone implies that electrical charges within the bone undergo displacement caused by the stress. We now know that the collagen in bone accounts for most of its piezoelectric properties.

Electrical activity observed in bone includes **stress-generated potentials** and **biopotentials (non-stress-generated potentials).** The original work in 1953 dealt with stress-generated potentials. These potentials do not depend on cell viability and arise from the motion of charges on collagen molecules that are fixed in the mineralized matrix. Dry tendon collagen is also piezoelectric. Bone stressed in bending develops a negative potential on the concave compression side and a relative electropositivity on the convex tension side. Thus, areas of compression are electronegative relative to areas of tension.

Steady voltages (biopotentials) also exist along intact living bone and in fracture callus. They are independent of stress, require viable cells, and are associated with metabolic activity. Electrodes placed along living bone demonstrate that the medullary canal is electronegative relative to the surface. The metaphysis is several millivolts more electronegative than the diaphysis, and the growth plate is an area of increased electronegativity, as is a fracture callus. Inactive areas of a bone are neutral or electropositive. These biopotentials are unaffected by denervation or the rate of blood flow.

Areas of electronegativity in bone coincide with sites of osteogenesis; areas of relative electropositivity (current entering) are associated with resorption. After a fracture, the injured area is electronegative. Recall that the direction of current flow is defined as the direction in which positive charge moves. In nonunion treatment, negative electrodes can be implanted in the fracture gap, and new bone forms around the negative electrode (cathode). It appears that one defect common to all nonunions is the loss of endogenous fracture electronegativity.

In summary, electronegative biopotentials are associated with osteogenesis at the growth plate and at the site of fracture healing. Stress-generated potentials arise from physical deformation of bone and are closely associated with remodeling. Areas of bone compression are electronegative relative to areas of tension. Charges on collagen molecules and electrokinetic effects are responsible for stress-generated potentials. Living cellular activity produces biopotentials.

FRACTURE HEALING

Bone is one of the few organs able to regenerate completely following an injury such as a fracture. Skin, kidney, brain, and other organs mostly repair by fibrous scar with little regeneration. Bones repair by bone formation and restoration of normal architecture through remodeling. This section discusses three types of fracture repair: (1) external bridging callus healing, (2) cancellous healing, and (3) primary bone healing.

External Bridging Callus Healing

A fractured long bone treated nonoperatively heals by producing an external bridging callus. Three overlapping morphological phases describe this healing: the inflammatory phase, the reparative phase, and the remodeling phase.

Inflammatory Phase

When a bone absorbs enough energy to fail, considerable tissue damage occurs. The bone itself fractures, the periosteum and endosteum tear, blood vessels rupture, and the surrounding muscle may be crushed and injured. Hematoma rapidly engulfs the area as a result of bleeding from the bone ends and from the torn vessels. Cells perish from the initial trauma, and many others die as a result of vascular disruption or thrombosis. For several millimeters from the fracture ends, cortical osteocytes die because of Haversian blood flow compromise.

The tissue damage elicits an acute inflammatory reaction. Local intact capillaries dilate, plasma exudes from the capillaries, and polymorphonuclear neutrophils, macrophages, and mast cells swarm into the area. The oxygen tension in the hematoma is very low, measuring about 6.25 mm Hg. Cellular metabolism in the hematoma is anaerobic, and the environment is acidic from the buildup of lactic and other organic acids. The hematoma coagulates into a fibrin clot, and macrophages begin the task of cleaning up cellular debris. Undifferentiated cells from the endosteum and cambium periosteal layer start to proliferate within the first few hours, and the process of osteoinduction begins.

Clinically, the extremity is swollen, erythematous, and exquisitely painful. This initial phase lasts for 3 to 4 days before the acute inflammatory reaction begins to subside. Removal of dead cells and tissue debris can last for several weeks, depending on the amount of initial tissue trauma.

Reparative Phase

The second, or reparative, phase is divided into the stage of **soft callus** and the stage of **hard callus.** The stage of soft callus begins as capillaries and fibroblasts invade and organize the clot into granulation tissue. Mesenchymal cells appear from the periosteum, endosteum, endothelium, and marrow. They differentiate into fibroblasts, chondroblasts, or osteoblasts. Nests of chondrocytes proliferate in the hypoxic depths of the clot, and the entire mass becomes a dense collage of fibrous and vascular tissue, fibrocartilage, and hyaline cartilage. Spicules of new bone form under the periosteum and in the more mature sections of the callus. Oxygen tension remains low, but the extracellular pH is more neutral. Cellular density increases until the bony ends are enveloped by a spindle-shaped, fibrous, and cartilaginous mass. The surface of the mass is strongly electronegative. The stage of soft callus usually lasts about 3 weeks and corresponds to the clinical time when the fracture is noted to be "sticky."

The major event during the stage of hard callus is conversion of the soft fibrous and cartilaginous callus into woven bone. Cartilage resorption proceeds in a fashion similar to endochondral bone formation but lacking the linear organization of the growth plate. Membranous bone formation is also apparent. The fusiform callus mass contains increasingly larger amounts of bone and becomes more rigid. The stage of hard callus usually lasts 3 to 4 months and corresponds to the clinical and radiographic determination of bony "union."

Remodeling Phase

Remodeling of the hard callus actually begins during the middle of the reparative phase and lasts for many years. The mechanisms of fracture remodeling are the same as normal bone remodeling but greatly accelerated. Lamellar bone replaces woven bone, new osteones replace old or damaged Haversian systems, the medullary canal is reestablished, and the diameter of the fracture callus is reduced by excess bone removal. Sensitive radioisotopic studies have detected remodeling of a fractured tibia as long as 7 years post-injury.

Cancellous Healing

Flat bones or those containing mostly cancellous bone heal by a different mechanism. Exuberant external bridging callus is not observed after fracture of bones such as the skull, a vertebral body, the patella, or the scapula because of the nature of cancellous healing. Less osteonecrosis follows fracture of cancellous bone because of the thin trabeculae and the close proximity of cancellous osteocytes to medullary capillaries. Furthermore, the injured trabeculae usually maintain intimate contact with each other following the fracture.

After the initial inflammatory phase, union of cancellous bone occurs largely by the process of **"creeping substitution."** Osteoblasts advance into the injured region and make new appositional bone directly on the surface of dead trabeculae. This healing of new bone over old is very fast, and union occurs in the order of 4 to 6 weeks after injury.

A special situation of cancellous healing occurs when the fracture site lacks a periosteum, as in the case of a femoral transcervical fracture, as reported by Kenzora and Glimcher. All bone cells proximal to the fracture die. The healing process begins with the ingrowth of a "syncytium" of vascular and undifferentiated cells from the distal end. Osteoblasts settle on the surface and deposit new woven bone on the dead trabeculae, giving the bone a radiodense appearance. Eventually, the woven and dead bone are replaced by new lamellar bone.

As new bone is just being deposited on dead subchondral trabeculae, extensive remodeling is proceeding distally in the vicinity of the fracture. If the articular cartilage and subchondral bone collapse prior to new bone deposition on the subchondral trabeculae, the articular cartilage loses its support, the femoral head becomes irregular, and degenerative arthritis ensues.

Primary Bone Healing

Fractures that are anatomically reduced and rigidly stabilized with compression plating unite by primary bone healing. Little if any external or internal callus appears. Healing is by direct growth of Haversian systems across the fracture site: BMU cutting cones burrow through cortical bone, cross the microscopic fracture gap, and enter cortical bone on the other side. Osteoblasts follow in the cutting cones' wake and weld the fracture with new lamellae. This continues until all dead bone is replaced and new Haversian systems are distributed according to the load characteristics of the bone. Strength returns at a slower rate than with external bridging callus, so caution is in order after removal of the internal fixation.

Nonunion

Nonunion occurs when healing stops prior to bony union. The resulting pseudarthrosis can be **atrophic** if the union is mostly fibrous or **hypertrophic** if cartilage predominates. Excess motion at the fracture site and tissue metaplasia produce a true **synovial pseudarthrosis.** The following factors interfere with healing and promote pseudarthrosis: open and comminuted fractures with massive soft tissue damage, iatrogenic interference such as repeated manipulations, extensive periosteal stripping, infection, poor blood supply, and soft tissue interposition. It is interesting that nonunions seem to be a uniquely human phenomenon, unheard of in wild animals and difficult to produce in laboratory animals without special immobilization techniques and other aggressive interventions.

BONE GRAFTS

The classic way to treat delayed union, nonunion, and skeletal defects is with a bone graft. There are four types of bone grafts: autograft, isograft, allograft, and xenograft. An **autograft** is the transfer of bone within the same person, such as the transfer of cancellous bone from iliac crest to the spine after Harrington instrumentation. An **isograft** is between identical twins, who share identical histocompatability antigens. **Allografts** are between genetically dissimilar members of the same species, and **xenografts** are from different species, such as implantation of bovine bone. Bone grafts can be used to stimulate healing or to replace lost tissue.

Bone grafts work because of three mechanisms. (1) A few surface cells survive the transplant and provide a source of osteoprogenitor cells. (2) Trabeculae act as a scaffolding for creeping substitution. This trellis function, also called **osteoconduction,** permits the ingrowth of capillaries and perivascular tissue, which in turn brings in osteoprogenitor cells to grow on the three-dimensional structure of the graft. (3) The graft is also a source of local growth factors such as bone morphogenetic protein. This is called **osteoinduction,** a process of recruitment of cells into the graft under the influence of diffusible proteins like BMP.

MYOSITIS OSSIFICANS

Posttraumatic skeletal muscle ossification (myositis ossificans) commonly affects the brachialis, the quadriceps, and the thigh adductors. Clinically, the site is warm, swollen, and tender, but the white cell count, erythrocyte sedimentation rate, and serum calcium and phosphate are normal. Alkaline phosphatase may be normal or abnormal (large lesions, particularly around the hip and pelvis, have enzyme elevations). A radiograph is abnormal by 3 to 4 weeks, showing soft tissue swelling and flocculated densities (like cotton candy). On close inspection a radiolucent zone can be seen to separate the lesion from the underlying periosteum. By 6 to 8 weeks the mass is sharply circumscribed, and by 5 to 6 months the bone is radiographically mature with obvious trabeculation. The radiograph and pathological sections show a zonal phenomenon of maturation. The innermost cells are proliferating undifferentiated mesenchymal cells, the middle zone is mostly osteoid, and the outer zone is maturing bone.

Management consists of rest during the early stages followed by active range of motion and progressive resistance exercises when pain and swelling subside. Surgical excision should be undertaken only after maturation to avoid recurrence.

GLOSSARY

Allograft Tissue transferred from one individual to other of the same species.
Autograft Tissue transferred within the same individual.
Basic multicellular unit (BMU) The unit of remodeling, a BMU is a cutting cone of osteoclasts and a wake of osteoblasts, capillaries, and undifferentiated cells that fill in the cavity with new lamellar bone.
Biopotential A non-stress-generated potential difference between two areas of bone caused by metabolic activity of living cells.
Canaliculi Small channels radiating from each lacunae through bone matrix, containing osteocytic cytoplasmic processes, providing a pathway for nutrient diffusion from the vessels in the Haversian canal to the osteocytes.
Cancellous bone Also called trabecular bone, found in the metaphysis of long bones and the center of others, it is a three-dimensional mesh of interlocking plates with high surface-to-volume ratio.

Circumferential lamellae Seams of new bone laid down on the surface and resembling layers of an onion.
Cortical bone Also called compact bone, the dense bone tissue making up the diaphysis of long bones and the cortex of all others, composed mostly of Haversian systems and grossly resembling ivory.
Cutting cone A group of osteoclasts that bore into bone creating a resorption cavity, followed by capillaries and osteoblasts constructing a new osteone.
External bridging callus The mechanism of healing of a long bone treated with cast immobilization, involving formation of a callus that goes through three phases: inflammatory, reparative, divided into the stage of soft callus and the stage of hard callus, and remodeling phase.
Growth factors Substances that modulate bone growth and are not nutrients; they increase synthesis of nucleic acids, protein, and cellular mass, either in osteoblasts or osteoclasts, depending on whether the factor is a stimulater of bone formation or bone resorption.
Haversian canal The central canal of a Haversian system (osteone) containing vessels and nerves around which the lamellae are arranged.
Haversian system Also called an osteone, the cylindrical functional unit of bone, built around a Haversian canal and bonded to neighbors by a cement line.
Howship's lacunae Resorption cavities, usually found on cancellous bone, produced by and containing osteoclasts.
Interstitial lamellae Remnants of previous osteones or circumferential lamellae that fill the gaps between complete osteones.
Isograft Transfer of tissue from one individual to another of identical genotype such as identical twins or inbred strains.
Lamellae Bone matrix and osteoblasts organized into layers.
Morphogenesis The first stage of cell activity in bone grafting or transplants, consisting of cell disaggregation, migration reaggregation, and proliferation; the morphogenetic phase is followed by the cytodifferentiation phase.
Osteoconduction A process initiated by bone grafts in which the graft stimulates the ingrowth of capillaries and osteoprogenitor cells from the recipient bed into the graft.
Osteoinduction The recruitment of undifferentiated cells into bone by agents like bone morphogenetic protein.
Osteone See Haversian system.
Primary bone healing Healing of a bone that is rigidly stabilized without the intervention of external or internal callus by direct extension of BMU activity across the fracture site.
Stress-generated potentials A strain-related, stress-induced electrical potential due to the displacement of fixed electrical charges, in the case of bone, due to collagen molecules in the mineralized matrix.

Volkmann's canals Channels running perpendicular or oblique to the long axis of bone, containing vessels that feed the haversian canals.

Wolff's law Proposed by Julius Wolff, professor at University of Berlin, in 1868 stating that every change in the form or function of a bone is followed by certain definite changes in its internal architecture and its external conformation according to mathematical laws.

Woven bone Immature or pathological bone tissue lacking an orderly arrangement of cells and collagen fibers.

Xenograft Transfer of tissue from one species to another of a different species.

BIBLIOGRAPHY

Borgens, R. B. (1984): Endogenous ionic currents traverse intact and damaged bone. *Science*, 225:478–482.

Brand, R. A. (1979): Fracture healing. In: *The Scientific Basis of Orthopaedics*, edited by J. A. Albright and R. A. Brand, pp. 289–311. Appleton-Century-Crofts, New York.

Castor, C. W., and Cabral, A. R. (1985): Growth factors in human disease: The realities, pitfalls, and promise. *Semin. Arthritis Rheum.*, 15:33–44.

Cruess, R. L. (1982): Physiology of bone formation and resorption. In: *The Musculoskeletal System: Embryology, Biochemistry, and Physiology*, edited by R. L. Cruess, pp. 219–252. Churchill Livingstone, New York.

Cruess, R. L. (1984): Healing of bone tendon, and ligament. In: *Fractures in Adults*, second ed., edited by C. A. Rockwood and D. P. Green, pp. 147–168. J. B. Lippincott, Philadelphia.

Friedenberg, Z. B., Harlow, M. C., Heppenstall, R. B., and Brighton, C. T. (1973): The cellular origin of bioelectric potentials in bone. *Calcif. Tissue Res.*, 13:53.

Fukada, E., and Yasuda, L. (1957): On the piezoelectric effect of bone. *J. Physiol. Soc. Jpn.* 12:1158–1162.

Glimcher, M. J., and Kenzora, J. E. (1979): The biology of osteonecrosis of the human femoral head and its clinical implications. I. Tissue biology. *Clin Orthop.*, 138:284–309.

Heppenstall, R. B. (1980): Fracture healing. In: *Fracture Treatment and Healing*, edited by R. B. Heppenstall, pp. 35–64. W. B. Saunders, Philadelphia.

McKibbin, B. (1978): The biology of fracture healing in long bones. *J. Bone Joint Surg.*, 60B:150–162.

Perren, S. M., and Rahn, B. A. (1978): Biomechanics of fracture healing. I. Historical review and mechanical aspects of internal fixation. *Orthop. Surv.*, 2:108–143.

Phillips, L. S., and Vassilopoulou-Sellin, R. (1980): Somatomedins. *N. Engl. J. Med.*, 302:371–380,438–446.

Prolo, D. J., and Rodrigo, J. J. (1985): Contemporary bone graft physiology and surgery. *Clin. Orthop.*, 200:322–342.

Raisz, L. G., and Kream, B. E. (1983): Regulation of bone formation. *N. Engl. J. Med.*, 309:29–35,83–89.

Treharne, R. W. (1981): Review of Wolff's law and its proposed means of operation. *Orthop. Rev.*, 10:35–47.

Urist, M. R., DeLange, R. J., and Finerman, G. A. M. (1983): Bone cell differentiation and growth factors. *Science*, 220:680–685.

6

Joints, Synovium, Articular Cartilage

JOINT MORPHOLOGY

Joints develop between the fourth and the eighth week of fetal life from interzonal mesenchyme (Chapter 3). Interzonal mesenchyme differentiates into fibrous tissue or fibrocartilage in nonsynovial joints; synovial mesenchyme forms in synovial joints. Selective cell degeneration gives rise to cleft formation and cavitation. By 10 to 12 weeks of fetal life, villous folds are apparent in the synovium. Movement of a synovial joint is essential for continued normal development.

Joints have many variations in geometry, structure, and function depending on motion and load-bearing requirements. In general, joints are divided into three types: fibrous, cartilaginous, and synovial (Table 6.1). **Synar-**

TABLE 6.1. *Classification of joints*

Type	Motion	Structure	Example
Fibrous	Immovable	Synarthrosis	(1) Suture: skull (2) Syndesmosis: distal tibiofibular (3) Gomphosis: tooth sockets
Cartilaginous	Slightly movable	Amphiarthrosis	(1) Synchondrosis: ischiopubic (2) Fibrocartilaginous: intervertebral disk (3) Symphysis: pubic symphysis
Synovial	Freely movable	Diarthroses	(1) Gliding: carpal joints (2) Ginglymus: interphalangeal (3) Condylar: knee (4) Ball and socket: hip (5) Ellipsoidal: radiocarpal (6) Saddle: thumb CMC

TABLE 6.2. *Joints containing fibrocartilage pads*

Knee	Medial and lateral meniscus
Wrist	Triangular fibrocartilage
Temperomandibular	Articular disk
Sternoclavicular	Articular disk
Acromioclavicular	Articular disk

throses are fixed, fibrous joints that have minimal movement. Examples include the skull sutures and the distal tibiofibular joint. **Amphiarthroses** are slightly movable cartilaginous joints such as the symphysis pubis and the intervertebral joints. **Diarthroses** such as the elbow, ankle, and knee are freely movable synovial joints enclosed by a tough, fibrous capsule.

In synovial joints, the bone ends are covered with articular cartilage and bathed in synovial fluid. This provides a low-friction surface that resists compression and shear forces generated during weight bearing and muscle action. The entire joint is confined by a joint capsule, which is perforated by vessels and nerves. Synovial membrane lines the inner surface of a capsule and covers most nonarticular parts of a diarthrosis.

Intra-articular structures can be extrasynovial or intrasynovial. For instance, the cruciate ligaments in the knee and the biceps tendon in the shoulder are intra-articular but extrasynovial. Synovial fluid does not contact these structures, and they receive nutrition from internal capillaries, not the synovial fluid. The menisci of the knee and other fibrocartilage pads are intra-articular and intrasynovial (Table 6.2). They add stability to the joint and act as biological washers to withstand rotation stresses.

An extra-articular area of special concern is the **enthesis,** the area where tendons, ligaments, or the joint capsule insert into the bone. The enthesis is a point of early calcification in diffuse idiopathic skeletal hyperostosis (DISH) syndrome (Fig. 6.1).

THE SYNOVIUM

The synovial membrane mediates the exchange of nutrients and waste between blood and joint fluid, synthesizes and secretes synovial fluid, and degrades particulate debris or worn-out macromolecules. The synovial membrane is actually not a true membrane in the anatomic sense, because it lacks a basement membrane and does not form a continuous barrier as do epithelia or endothelia. Nonetheless, many authors continue to refer to the synovium as the synovial membrane.

The synovium is composed of two histological layers: (1) the **intima** and (2) the **subintima.** The synovial intima is a compact layer of **synovial lining cells** that have few points of contact and no desmosomes. The intima is one to four cells deep, and it is smooth, pink and glistening to the naked eye.

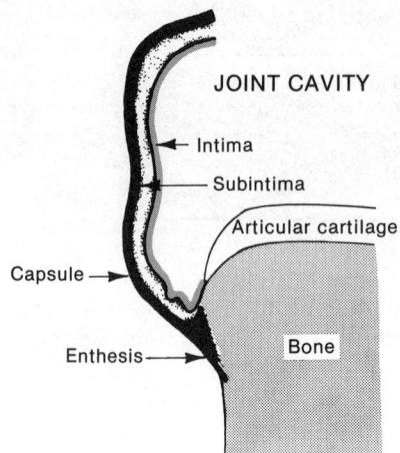

FIG. 6.1. General synovial joint structure. Intima lines the inside of the joint cavity, and it sits on a bare area of bone prior to ending at the articular cartilage. The area of insertion of capsule, ligaments, and tendons into bone is called the enthesis. Modified from Dieppe and Calvert (1983).

The synovium is thrown into small delicate folds or villi at the articular cartilage margins (areas that allow stretching with joint motion). Synovial lining cells are enmeshed in a loose matrix of collagen fibers and anchored to the underlying subintima by fibronectin and other extracellular proteins.

Subintima is a highly vascular connective tissue layer classified as either **fibrous, fatty,** or **areolar** based on the major connective tissue component. Subintima contains lymphatics, blood vessels, fenestrated capillaries, and nerves. As a rule, a joint is innervated by branches of all nerves that supply muscles acting across the joint (Hilton's law).

The intima is composed of three types of synovial lining cells (Table 6.3), classified according to ultrastructure as **type A** (macrophage-like), **type B** (synthetic–secretory), and **type C** (intermediate). Type A cells comprise about 60 percent of the synovial lining cells. Their ultrastructure reflects an aggressive phagocytic ability, showing a prominent Golgi, numerous vacuoles, filopodia, but sparse endoplasmic reticulum. Thirty percent of the synovial lining cells are type B cells, and 10 percent are type C. Type B cells have abundant rough endoplasmic reticulum to support protein and hyaluronic acid synthesis, and type C cells are intermediate in structure with both phagocytic and synthetic subcellular structure.

TABLE 6.3. *Synovial Lining Cells*

Type	Major function	Percentage of population
A	Phagocytic	60%
B	Secretory	30%
C	Intermediate	10%

Shape and volume of the joint space changes as the joint goes through its range of motion, and the synovium accommodates to this change by stretching and eliminating slack in the villous areas of joint recesses. The capsule and surrounding ligaments are often the limiting factor in maximum motion. The capsule merges with the periosteum and bone at the enthesis, leaving a small area of bare bone between the enthesis and the edge of the articular cartilage. The synovium continues over this bare bone area to end at the margin of articular cartilage. These bare areas of juxtiarticular bone covered only by synovium undergo early destruction by proliferative synovium in rheumatoid and hemophilic arthritis, producing subchondral erosion and cysts.

SYNOVIAL FLUID

The term synovia was coined by Paracelsus Bombastus of Hobenheim (1493–1541) for joint fluid because it resembled egg white, and in fact normal synovial fluid is clear, pale yellow and viscous. It is a filtrate of plasma containing the same electrolytes and small molecules as plasma but a lower protein concentration (usually 1.5–2 grams/deciliter, mostly albumin). Proteins of molecular weight greater than 150,000 such as fibrinogen or the complement cascade are retained within subsynovial capillaries; thus, normal synovial fluid does not clot. Proteoglycans, enzymes, and hyaluronic acid are added to the fluid by type B synovial lining cells. Hyaluronic acid, a glycosaminoglycan, is responsible for most of the viscosity of joint fluid. Synovial fluid hyaluronic acid, along with other glycoproteins, such as lubricin, lubricates the joint surfaces and provides a vehicle for nutrition to the chondrocytes.

A high lubrication efficiency is apparent from the friction coefficient of synovial joints, which is in the order of 0.003 to 0.015. Recall that the coefficient of friction is the ratio of the force necessary to start one body moving over another body to the load pressing them together. The friction coefficient of a road tire is 1, and that of an ice skate on ice is about 0.02. Synovial fluid demonstrates physical properties of viscosity, elasticity, and plasticity. It displays **non-Newtonian properties.** That is, with low shear rates, the fluid is highly viscous, but with high shear rates, the viscosity drops (Newtonian behavior means that viscosity is independent of shear rates).

William Hunter observed in 1743 that adult articular cartilage contained no blood vessels. Recent investigators have shown that the subchondral bone plate blocks significant transport of nutrients from underlying bone to articular cartilage. Thus, articular chondrocytes absolutely rely on synovial fluid for nourishment and for clearance of metabolites. Glucose, calcium, and small ions diffuse freely from synovial fluid into cartilage. Intermittent compression of cartilage by normal movement acts as a pump mechanism to facilitate movement of all molecules in and out.

Synovial Fluid Analysis

Normal synovial fluid is clear, transparent, and pale yellow. Viscosity is high because of the concentration and high molecular weight of hyaluronic acid (a glycosaminoglycan copolymer of N-acetylglucosamine and glucuronate). Glucose is almost equal to the serum concentration, and total protein is about 1.8 grams per deciliter, mostly albumin with minimal globulin and no fibrinogen. Synovial fluid is relatively acellular with ten to 100 cells per cubic millimeter, mainly monocytes with less than 25 percent polymorphonuclear neutrophils. The mucin clot is excellent (mucin clot refers to the precipitation of a hyaluronate–protein complex from synovial fluid when acetic acid is added).

Joint effusions can be classified into four categories: (1) noninflammatory, (2) inflammatory, (3) infectious, and (4) hemorrhagic (Table 6.4).

Noninflammatory joint effusions occur most often in osteoarthritis or following significant intra-articular trauma such as meniscal tears. The fluid is clear (the ultimate test of clarity being the ability to read print through a glass tube filled with the fluid). Glucose, protein, and cell count are normal. Viscosity is also normal, and the mucin clot is excellent.

Inflammatory effusions result from diseases such as rheumatoid arthritis, connective tissue disease, or crystal-induced synovitis. The fluid is opaque from cells and cellular debris. Glucose is low, and protein is high (including immunoglobulins, micro- and macroglobulins, and complement). Rice bodies, a conglomerate of fibrin, fibronectin, and collagen, are frequently present in chronic rheumatoid effusions. Viscosity is low, the mucin clot is poor, and the white cell count can vary from 2,000 to 75,000, mainly neutrophils.

Microscopic examination using polarizing lenses permits identification of crystals in synovial fluid (monosodium urate in gout or calcium pyrophosphate in pseudogout). Two perpendicular polarizing lenses will block the passage of light, but a birefringent crystal like monosodium urate will bend the light to permit passage through the ocular lens. Monosodium urate crystals are yellow when oriented parallel to the axis of the compensator (a mnemonic is YUP: yellow–urate–parallel). Calcium pyrophosphate is blue when parallel.

Infectious synovial effusions have a thick, purulent appearance and are quite opaque. Glucose concentration is very low, often 10 milligrams per deciliter or less. As a general rule, a glucose concentration less than half that of the blood glucose suggests bacterial infection. In a septic joint, cell count is high, 50,000 to 200,000 and 75 percent of the cells are polymorphonuclear neutrophils. Cell counts in tuberculous arthritis, gonococcal arthritis, and partially treated pyogenic infections can produce lower cell counts. The Gram stain is occasionally diagnostic, and the culture is usually positive if antibiotics have not been started.

TABLE 6.4. Synovial fluid analysis

	Normal	Noninflammatory	Inflammatory	Infectious	Hemorrhagic
Clarity	Transparent	Transparent	Cloudy to opaque	Opaque	Opaque
Color	Clear to yellow	Yellow	Yellow to green	Yellow to green	Pink to red
Viscosity	High	High	Low	Variable	Low
Cells per cubic millimeter	20–200	200–2,000	2,000–75,000	50,000–200,000	Variable
Polys	<25%	<25%	>50%	>75%	<25%
Glucose	<Serum	<Serum	Lower	Low	<Serum
Protein (grams per deciliter)	<2.5	<2.5	>2.5	>2.5	<2.5
Examples		(1) DJD	(1) RA	(1) Bacterial	(1) ACL tear
		(2) Meniscus tear	(2) Gout	(2) Fungal	(2) Osteochondral fracture
		(3) Osteochondritis dissecans	(3) Rheumatic fever	(3) TB	(3) PVNS

Hemorrhagic effusions may occur in systemic conditions such as hemophilia or overanticoagulation and also in focal conditions such as villonodular synovitis, osteochondral fracture, or anterior cruciate ligament rupture. Joint fluid is opaque, but the color depends on the age of the effusion: red or pink early on, yellow or greenish in an old hemorrhagic effusion.

CARTILAGE VARIETIES

Cartilages are made of three tissue elements: (1) chondrocotes, (2) amorphous matrix, and (3) various fibers. The tissue as a whole has a low metabolic rate and a high matrix-to-cell ratio. Variations in matrix and fiber composition determine the physical properties of cartilage. Tensile and shear strength come from the presence of type II collagen. Compressibility or resilience comes from the proteoglycans and other glycoproteins. Mature cartilages are avascular, aneural, and alymphatic. Three major types of cartilage are recognized: (1) hyaline cartilage, (2) fibrocartilage, and (3) elastic cartilage (Table 6.5).

Hyaline cartilage is a descriptive term for the most abundant cartilages, all having a glassy, translucent, bluish-pearly appearance (Gr. *hyalos,* glass). Subtypes of hyaline cartilage include fetal cartilages, articular cartilages, epiphyseal plates, and respiratory and costal cartilages. The matrix is 70 to 80 percent water and typically basophilic because of the cationic staining of the glycosaminoglycan (Table 6.5). Chondrocytes occupy only 5 percent of the volume of hyaline cartilage and occur as isolated cells or as cell nests (also called isogenous groups) in lacunae.

The matrix is called territorial in the pericellular regions and interterritorial in the remainder. The territorial matrix stains more intensely basophilic (deeper blue in hematoxylin and eosin stains) than the interterritorial matrix, presumably because of the increased concentration of glycosaminoglycan.

TABLE 6.5. *Approximate composition of the various types of cartilage*

Cartilage	Water (%)	Solids (%)			
		Collagen	GAG	Elastin	Other[a]
Articular	72	66	18	—	16
Epiphyseal	81	37	15	—	48
Fibrocartilage	74	78	2	0.6	19
Elastic	71	53	12	19	16

[a]Includes monocollagen proteins, calcium phosphorus, other ions, and macromolecules such as DNA and RNA.
From Stockwell (1979).

Fibrocartilage is composed of scattered chondrocytes in a dense white matrix of collagen fiber bundles. The collagen content is mainly type I and type III with little if any type II. The proteoglycan content is lower than that of hyaline cartilage. Fibrocartilage is found in the intervertebral disks, the symphysis pubis, menisci, the acetabular labra, and as a regenerated substitute for articular cartilage. The fibrocartilages vary somewhat in percentage of collagen subtypes and glycosaminoglycan content.

Elastic cartilage contains typical chondrocytes, but the matrix includes a network of yellow elastic fibers as well as collagen fibers and proteoglycans. The ears, epiglottis, and corniculate cartilages of the larynx contain elastic cartilage.

ARTICULAR CARTILAGE

Articular cartilage is a type of hyaline cartilage specialized for (1) low-friction movement, (2) load distribution and transmission, and (3) shock absorption. It varies in thickness form 2 to 7 millimeters depending on the joint in question (very thick on the patella, thin on the metacarpals). Unlike other hyaline cartilages, articular cartilage is not covered by perichondrium. The surface appears smooth to the naked eye, but scanning electron microscopy shows small pits, elevations, and ridges, important for lubrication and nutrition.

Articular cartilage does not have a homogeneous structure. Both chondrocyte morphology and collagen fiber size and orientation vary from surface to subchondral bone, giving rise to a zonal pattern. The four **zones** are (I) **tangential** zone, (II) **transitional** zone, (III) **radial** zone, and (IV) **calcified** zone. The **tidemark** separates the radial zone from the calcified cartilage zone (Fig. 6.2).

Zone I, the **tangential** zone, is most superficial and in immediate contact with the synovial fluid. Chondrocytes are flattened or ovoid and lie parallel to the articular surface. Small collagen fibers run throughout the matrix and align in a predominately parallel orientation to the surface.

The most superficial layer of the tangential zone contains a very low concentration of proteoglycan and thus does not stain with cationic dyes, prompting MacConaill to call it the lamina splendens. It is actually not a separate lamina or layer at all but simply an area high in collagen and low in proteoglycan, able to resist high shear and compressive forces where the opposing joint surfaces touch.

Zone II is the **transitional** or intermediate zone. Chondrocytes are round and plump, and they have a random, haphazard dispersion within the matrix. Proteoglycan concentration is high, and collagen fibers cross and decussate as they pursue oblique courses throughout this zone.

Zone III is the **radial** zone, which occupies the area between the transi-

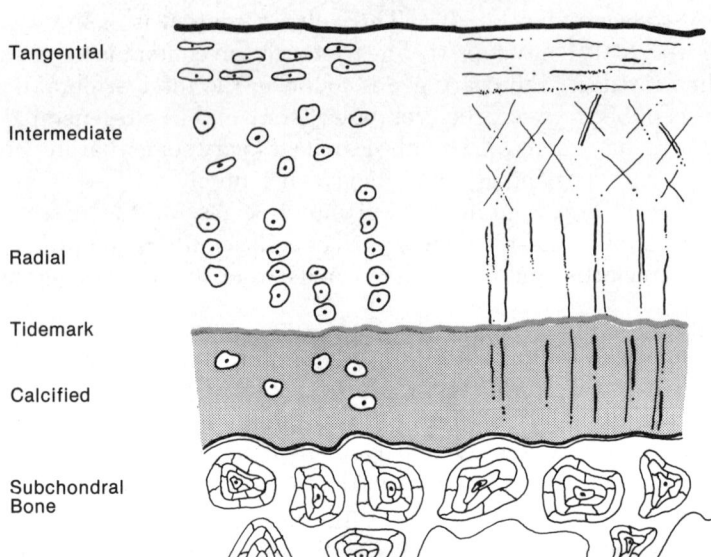

FIG. 6.2. Schematic representation of the cell orientation and collagen fibril orientation in the four histologic layers of articular cartilage. Chondrocytes are flattened, and collagen fibrils are mostly parallel to the surface in the tangential zone. Fibrils decussate in the intermediate zone and lie more perpendicular to the surface in the radial zone and in the calcified cartilage.

tional zone and the tidemark. Chondrocytes are large and round and tend to line up in rows several cells high that are perpendicular to the surface. Collagen fibers are thick and numerous with a radial orientation that is also perpendicular to the surface.

The **tidemark** is a wavy basophilic line 2 to 5 micrometers thick that clearly demarcates the uncalcified from the calcified cartilage. The exact chemical composition and structural function of the tidemark are obscure.

Zone IV is the **calcified** or deep zone that lies adjacent to subchondral bone. Chondrocytes are small and variable in size and histological integrity, some cells appearing degenerated or necrotic. The matrix is impregnated with calcium salts, and collagen fibers have a more random orientation. The calcified cartilage firmly interlocks with the undulating surface of underlying subchondral bone.

Microscopic studies of articular cartilage demonstrate collagen fibers originating from subchondral bone. They extend through the calcified and radial zones, ending or crossing in the tangential zone. Benninghoff noted this arrangement in 1925 and suggested that the fibers actually formed arcades to withstand compressive forces. However, it is now known that the arcades are not discrete collagen bundles but collective orientations of many fibers, mostly perpendicular to the surface in the radial zone, crossing or decussating in the transitional zone, and parallel to the surface in the tangential zone.

Metabolic activity of articular cartilage is low, but internal remodeling constantly occurs. Some proteoglycans and glycoproteins turn over at a fairly rapid rate, with a half-life of 5 to 8 days, but the bulk of the proteoglycan is more stable, with a half-life over 800 days. The half-life of collagen fibers of a normal adult joint is much longer, probably on the order of years.

Aging of Articular Cartilage

Young cartilage is smooth, pinkish-white, and glistening, whereas old cartilage is thin, firm, yellowish-opaque, and somewhat brittle. As chondrocytes age, they synthesize less chondroitin-6-sulfate and progressively more chondroitin-4-sulfate and keratan sulfate. Young chondrocytes make proteoglycans with long chondroitin sulfate chains and short keratan sulfates. With age, the length of chondroitin sulfate chains decreases, and keratan sulfate length increases. Thus, both chain length and composition of glycosaminoglycans change with age. This means that proteoglycans from older individuals bind less water because of the smaller chondroitin sulfates, and cartilage stiffness increases as a result of increased content of keratan sulfate. Collagen seems to change little, although some evidence suggests that the amount of cross linking may increase.

Chondromalacia

The term chondromalacia means "soft cartilage" and refers to abnormal cartilage that feels soft as opposed to normal cartilage, which feels firm and elastic. The term chondromalacia should be used only to describe demonstrated pathological changes in articular cartilage and not the clinical syndrome of retropatellar pain or crepitus.

Although chondromalacia can affect any articular surface, the patella is most frequently involved. A partial list of etiological considerations includes (1) malalignment problems (increased quadriceps angle, patella alta, vastus medialis insufficiency, excess lateral pressure), (2) trauma (direct impact, joint dislocation or subluxation), (3) disuse (immobilization effects, lack of intermittent loading as occurs at the odd facet of the patella).

Early pathological changes in chondromalacia include a limited area of cartilage softening and loss of turgor. At arthroscopy, a probe detects increased sponginess and loss of firmness. Microscopy shows loss of orien-

TABLE 6.6. *Chondromalacia grading systems*

		Goodfellow et al. (1976)	
Grade	Outerbridge (1961): chondromalacia	Basal degeneration	Superficial degeneration
Grade I	Softening and swelling	Softening	Fibrillation
Grade II	Fissure, fragmentation <1.3 centimeters diameter	Blister formation	Fissure formation
Grade III	Fissure, fragmentation >1.3 centimeters diameter	Crabmeat ulceration	Fragmentation
Grade IV	Erosion to subchondral bone	Crater and eburnation	Crater and eburnation

tation and cohesion between collagen fibers in the superficial tangential zone. With progression, the superficial cartilage layers separate from the deeper layers, producing a cystic space that is noted as a localized bulge (blister) on the surface. Rupture of the blister initiates a sequence of ulceration, collagen fiber fibrillation, fissure formation, and fragmentation. Many cartilage fragments remain attached to the subchondral bone and have a "crabmeat" appearance. If the process continues, subchondral bone is eventually exposed. The lesion eventually progresses to a crater of eburnated bone surrounded by a rim of frayed and fragmented cartilage.

Goodfellow, Hungerford, and Wood noted two type of chondromalacia: **basal degeneration,** a localized type caused by trauma, and **superficial degeneration,** a nontraumatic lesion. Basal degeneration begins with softening and progresses to blister formation, crabmeat ulceration, and finally crater formation. Superficial degeneration begins with surface fibrillation and progresses to fissure formation, fragmentation, and finally eburnation. Chondromalacia may progress to osteoarthritis in some cases, but a definite causal relationship has not been firmly established. Table 6.6 lists the chondromalacia grading systems of Outerbridge (1961) and Goodfellow et al. (1976).

Osteoarthritis

The most prevalent abnormality of articular cartilage is osteoarthritis (OA). Eighty-five percent of people between the ages of 70 and 79 years have radiographically observable OA, and 80 percent of those over 55 have clinical evidence of OA in at least one joint (most commonly the knee). Epidemiologic studies reveal no significant geographic or ethnic patterns to OA, although a strong relationship to obesity has been shown. The popular name for OA in the American literature is "degenerative joint disease," but this is actually a poor and incorrect term because the biochemistry and cell biology of OA are predominately anabolic, synthetic, and inflammatory, not degenerative.

Osteoarthritis can be classified into two categories: (1) primary (idiopathic) and (2) secondary. Primary OA occurs in people with no history of joint injury or joint-associated disease, and primary OA can be localized or generalized. Secondary OA occurs after previous trauma or after inflammatory, metabolic, or other joint disease (e.g., ochronosis, gout, hemophilia).

Osteoarthritis is primarily a disease of cartilage and starts as a focal erosive lesion of articular cartilage. The normally smooth articular surface becomes fibrillated and marked with fissures and clefts, eventually exposing the subchondral bone. The earliest biochemical change in the fibrillated cartilage is an increase in water content (coinciding with a decrease in safranin O staining). A depletion in both glycosaminoglycan concentration and proteoglycan aggregation is apparent. Chondrocytes, which normally maintain a perfect balance of matrix synthesis and degradation, now have a two- to fourfold increased synthesis of type II collagen and proteoglycans, apparently attempting to reverse the matrix depletion. Furthermore, the chondrocytes, normally resting cells, proliferate by cell division to form clones of cells scattered throughout the matrix.

The increased synthesis also involves the subchondral bone. Osteoblasts increase the thickness and stiffness of the subchondral plate. As the chondrocyte proliferation progresses, joint surfaces eventually become incongruous, and marginal osteophytes extend the cartilage boundaries (apparently a natural attempt to decrease force per unit area). Tendons, ligaments, and capsule around a joint with OA also hypertrophy and enlarge. Thus, OA cartilage is metabolically active, synthetic, and switched on, not a passive, wear-and-tear deterioration (Table 6.7).

Products of cartilage destruction (collagen fragments, cell debris) spill into

TABLE 6.7. *Characteristics of osteoarthritis: Increased metabolism with articular compromise*

Joint component	Change
Matrix	Increased water content
	Decreased proteoglycan concentration and aggregation
Chondrocytes	Increased proteoglycan synthesis
	Cell division, cloning
Cartilage	Fibrillation, fissuring, peripheral growth
Synovium	Inflammation
Bone	Eburnation, osteophytes, subchondral sclerosis, cysts
Ligament, capsule	Hypertrophy

the joint fluid and are phagocytosed by synovial macrophages, causing release of inflammatory mediators. Inflammation results in further cartilage destruction, which perpetuates the inflammatory reaction. Joint fluid increases with synovial proliferation, and penetration of the fluid through subchondral microfracture sites contributes to cyst formation.

In OA, synthetic reactions produce extensive remodeling of the entire bone ends and the surrounding connective tissues. A radiograph of osteoarthritis shows sclerosis and hypertrophic new bone at joint margins and a loss of the joint space as a result of cartilage destruction. The main clinical symptom is moderately intense pain, and the most sensitive sign is the grating sensation, crepitus, that can be felt during joint motion.

The exact cause of OA is unknown, but theories include endocrine, genetic, traumatic, and biomechanical causes. Radin proposed one of the leading biomechanical theories in which the primary event begins in the subchondral bone. Repetitive impulse loading generates subchondral microfractures that heal with trabecular callus formation. More microfractures mean more callus, and the bone gradually becomes less compliant. This increased subchondral stiffness (loss of compliance) decreases shock absorption, and greater stresses are transferred to the matrix and to the chondrocytes. The matrix deteriorates, and the chondrocytes switch into high gear to restore the lost matrix.

Biochemical and cellular events of OA may continue until the joint is completely destroyed. However, articular cartilage is a dynamic tissue, and if abnormal joint mechanics are eliminated (such as following realignment osteotomies), these biochemical and cellular events of OA may halt. If the mechanics are corrected early enough, some of the destructive changes may even reverse, as evidence by increased joint space and decreased osteophyte size following appropriate osteotomies. But in many situations, management focuses on control of inflammation and symptomatic elimination of pain.

Rheumatoid Arthritis

Rheumatoid arthritis (RA) is a chronic systemic inflammatory disorder that progressively destroys synovial joints. It affects about 1 percent of all adults, women two to three times more often than men. Much is known about the pathogenesis of RA, but the etiology remains obscure despite a plethora of autoimmune, metabolic, infectious, inflammatory, psychosocial, and genetic theories.

Extra-articular manifestations of RA include arteritis, lymphadenopathy, neuropathy, pericarditis, and uveitis. Serum and joint fluids contain rheumatoid factors (antibodies to the Fc portion of the IgG molecule). Most rheumatoid factors are IgM, but a small minority are IgG and IgA.

The initial joint lesion begins in the synovium with microvascular injury, edema, and a primary inflammatory reaction. Synovial lining cells undergo hyperplasia, and hypertrophy, reaching a depth of six to ten cells, and edematous villi protrude into the joint. Immune complexes form in the synovium and activate the complement cascade. T lymphocytes, plasma cells and macrophages invade and fill the subintimal connective tissue stroma. Neutrophils phagocytize the immune complexes and other cellular debris, releasing lysosomal enzymes and lymphokines into the synovium and stimulating prostaglandin synthesis. As these reactions continue, the synovium enlarges and grows as a pannus over the articular cartilage and intraarticular structures.

Pannus is a mass of hypertrophic synovium containing capillaries, fibroblasts, and inflammatory cells with all the enzymes necessary to dissolve cartilage, bone, ligaments, and tendons. Collagenase, neutral proteases, hyaluronidase, and proteoglycanases degrade cartilage matrix around the pannus. These enzymes spill into synovial fluid, thus continuously bathing the joint surfaces with degradative enzymes. Chondrocytes also share in the destruction of cartilage. Synovium releases a factor called **catabolin** that switches chondrocyte metabolism from synthetic to matrix-degradative enzymology. The end result is synovial and capsular scarring, loss of articular cartilage, periarticular osteoporosis, and, in some cases, fibrous or bony ankylosis. The enzymatic softening and deterioration of articular structures coupled with weight-bearing or daily muscular forces produce the characteristic deformities of RA: swan neck and boutonniere deformities, ulnar drift, arthrokatadesis, Baker's cyst, etc.

Articular Cartilage Injury

Mankin (1982) notes that articular cartilage has a unique response to injury because cartilage is avascular and alymphatic. Furthermore, unlike most cells of the body, chondrocytes are relatively insensitive to hypoxia. The classic tissue response to injury consisting of necrosis, inflammation, and repair is absent. The actual response depends on the nature of the injury, whether it is (1) a superficial laceration, (2) a deep laceration into the subchondral bone, or (3) a crush or impaction.

A **superficial laceration** of articular cartilage (one above the tidemark) is a permanent injury. In the absence of blood vessels, there is no local bleeding, hematoma, fibrin clot, or granulation tissue. The inflammatory phase of healing is totally absent. A few chondrocytes next to the laceration will divide and synthesize new matrix components, but this short-lived burst of activity is over by 2 weeks, and no further healing occurs. From recent experimental work, a superficial laceration appears to remain completely unchanged. No healing occurs, but the laceration does not usually progress to either chondromalacia or osteoarthritis. It is a static, permanent lesion.

A **deep cartilage laceration** that penetrates the subchondral bone will definitely damage blood vessels. Local bleeding produces a hematoma that becomes organized and invaded by granulation tissue. This is a dynamic lesion, and the laceration defect fills with fibrovascular tissue, which eventually matures into fibrocartilage. Salter (1980) discovered that deep cartilage lacerations heal more rapidly and with a tissue more closely resembling hyaline cartilage if the injured joint is subjected to continuous passive motion. The exact mechanism for this motion metaplasia is unknown, but it holds considerable promise for practical clinical applications.

The response of articular cartilage to blunt trauma depends on the magnitude of the impact. Repo and Finlay (1977) reported that cartilage impacted at 10 percent strain or less survived without injury, but impaction at a strain of 40 percent produced chondrocyte death. Surface dents and defects were noted after high-energy impaction, indicating failure of the collagen network as well as cellular death. These also are permanent injuries.

GLOSSARY

Amphiarthrosis A slightly movable, cartilaginous joint such as the symphysis pubis, the ischiopubic synchondrosis, or the intervertebral joints.

Catabolin A putative factor that switches chondrocyte metabolism from anabolic–synthetic to catabolic–degradative.

Chondromalacia A pathological diagnosis of focal softened, fibrillated, or fissured articular cartilage.

Coefficient of friction The ratio of the force necessary to start one body moving over another body to the load pressing them together.

Diarthrosis A freely movable synovial joint such as the hip, knee, ankle.

Elastic cartilage A specialized cartilage containing elastin plus collagen and proteoglycan, found in the ears and the epiglottis.

Enthesis That point where a tendon, ligament, or joint capsule attaches to and inserts into bone.

Fibrocartilage Tough, dense cartilage making up menisci, intervertebral disks, the acetabular labrum, and healed articular cartilage, containing type I and type III cartilage in a proteoglycan matrix.

Hyaline cartilage The most abundant cartilage, consisting mostly of type II collagen in a chondroitin-sulfate-rich proteoglycan matrix, subtypes of which include articular cartilage, epiphyseal cartilage, respiratory and costal cartilages.

Lubricin A glycoprotein found in synovial fluid that helps reduce the coefficient of friction of articular cartilage.

Mucin clot A precipitate of hyaluronic acid and protein that forms in normal synovial fluid after addition of acetic acid.

Newtonian fluid A fluid in which the viscosity is independent of shear rates, such as water.

Non-Newtonian fluid A fluid in which the viscosity varies with shear rate, such as synovial fluid, which has a high viscosity at low shear rates but a low viscosity at high shear rates.

Pannus An expanding mass of hypertrophic synovium capable of enzymatically dissolving most connective tissue elements.

Rheumatoid factors Antibodies, mostly IgM, with antigenic determinants to the Fc portion of the IgG molecule, i.e., anti-immunoglobulin antibodies.

Synarthrosis An immovable, fibrous joint such as the distal tibiofibular syndesmosis, skull sutures, or tooth sockets.

Synovial lining cells Those cells making up the synovial intima and consisting of type A (phagocytic), type B (secretory), and type C (intermediate) cells.

Synovium The inner lining of a diarthrosis consisting of two histological layers, the intima (synovial lining cells) and the subintima.

Tidemark A wavy, basophilic line in articular cartilage that divides the radial zone from the calcified cartilage zone.

BIBLIOGRAPHY

Benson, D. R., Castles, J. J., Wolf, A. W., Shapiro, R. F., and Riggins, R. S. (1981): Synovial fluid analysis in the diagnosis of joint diseases. *Contemp. Orthop.*, 3:430–440.

Bland, J. H., and Cooper, S. M. (1984): Osteoarthritis: A review of the cell biology involved and evidence for reversibility. Management rationally related to known genesis and pathophysiology. *Semin. Arthritis Rheum.*, 14:106–133.

Cox, J. (1983): Chondromalacia of the patella: A review and update. *Contemp. Orthop.*, 6:17–30, 35–46.

Dieppe, P., and Calvert, P. (1983): *Crystals and Joint Disease*. Chapman and Hall, New York.

Ghadially, F. N. (1981): Structure and function of articular cartilage. *Clin. Rheum. Dis.*, 7:3–28.

Ghadially, F. N. (1983): *Fine Structure of Synovial Joints*. Butterworths, London.

Goodfellow, J., Hungerford, D. S., and Woods, C. (1976): Patello-femoral joint mechanics and pathology. 2. Chondromalacia patellae. *J. Bone Joint Surg.*, 58B:291–299.

Harris, E. D. (1984): Pathogenesis of rheumatoid arthritis. *Clin. Orthop.*, 182:14–22.

Hasselbacher, P. (1981): Structure of the synovial membrane. *Clin. Rheum. Dis.*, 7:57–66.

Henderson, B., and Pettipher, E. R. (1985): The synovial lining cell: Biology and pathobiology. *Semin. Arthritis Rheum.*, 15:1–32.

Mankin, H. J. (1982): The response of articular cartilage to mechanical injury. *J. Bone Joint Surg.*, 64A:460–466.

Outerbridge, R. E. (1961): The etiology of chondromalacia patella. *J. Bone Joint Surg.*, 43B:752–757.

Repo, R. U., and Finlay, J. B. (1977): Survival of articular cartilage after controlled impact. *J. Bone Joint Surg.*, 59A:1068–1076.

Ryu, J., Treadwell, B. V., and Mankin, H. J. (1984): Biochemical and metabolic abnormalities in normal and osteoarthritic human articular cartilage. *Arthritis Rheum.*, 27:49–57.

Salter, R. B., Simmonds, D. F., Malcolm, B. W., Rumble, D. J., McMichael, D., and Clements, N. D. (1980): The biological effect of continuous passive motion on healing of full-thickness defects in articular cartilage. An experimental investigation in the rabbit. *J. Bone Joint Surg.*, 62A:1232–1251.

Shahriaree, H. (1985): Chondromalacia. *Contemp Orthop.*, 11:27–39.

Stockwell, R. A. (1979): *Biology of Cartilage Cells*. Cambridge University Press, Cambridge.

7

Nerve and Muscle

THE NEURON

Two general categories of cells comprise the central and peripheral nervous system, (1) the electrically excitable neurons and (2) the nonexcitable neuroglia, ependymoma, and Schwann cells.

Neurons are highly elongated cells specialized to receive, conduct, and transmit electrical signals called action potentials. They have diverse sizes and shapes, but all neurons have four morphological parts: (1) a **cell body** (perikaryon), which is the electrical integration and biosynthetic center, containing a nucleus and most subcellular organelles; (2) **dendrites,** which extend antenna-like from the cell body to increase the surface area for signal (synaptic) input; (3) an **axon,** which conducts action potentials away from the cell body; and (4) **synapses,** which are the axonal termini where neurons communicate with other cells (Fig. 7.1.)

A large, round, pale nucleus occupies much of the neuronal cell body, indicating active RNA synthesis and processing. The cell body also contains basophilic **Nissl bodies,** aggregates of endoplasmic reticulum, polyribo-

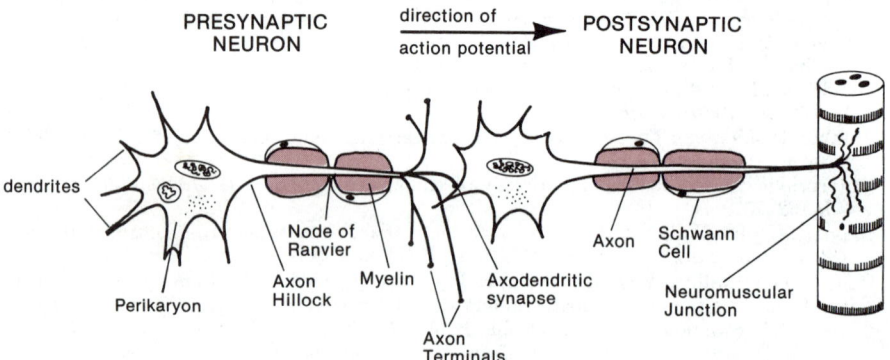

FIG. 7.1. Neuron structure. Action potentials are unidirectional, from the presynaptic neuron, through the synapse, to the postsynaptic neuron.

somes, unbound ribosomes, and RNA (areas of active protein synthesis). Mitochondria and a large Golgi complex energize and package neurotransmitters and proteins. Lysosomes and pigment granules such as lipofuscin are also present.

Axons arise from a pyramidal base on the cell body called the axon hillock. Motor axons conduct action potentials peripherally (efferent), and sensory axons conduct potentials centripetally (afferent) towards the CNS. Proteins and neurotransmitters synthesized in the cell body are transported down the axon and also out into the dendrites. Two types of axonal transport can be detected, slow and fast. Slow transport is the bulk flow of axoplasm (including vesicles, mitochondria, and lysosomes) at a rate of 1 to 3 millimeters per day. Rapid transport is a selective movement of special molecules and proteins at a rate of 100 millimeters per day (range 50–2,000). The mechanisms of slow and fast transport are incompletely understood, but microtubules, microfilaments, actin, and ATP appear to be involved.

Glial cells provide mechanical and metabolic support to neurons. Central nervous system glial cells include astrocytes, oligodendrocytes, ependymal cells, and microglial cells. The peripheral nervous system has only one type of glial cell, the **Schwann cell.** As a nerve root exits the spinal cord and becomes a peripheral nerve, Schwann cells replace oligodendrocytes as supporting glia.

Peripheral nerve fibers are either nonmyelinated or myelinated. **Nonmyelinated axons** such as pain and olfactory fibers have a spare protection by being wrapped in single longitudinal invaginations of Schwann cell membrane. **Myelinated axons** are completely insulated from surrounding tissue by a myelin sheath, which consists of up to 100 regularly spaced concentric layers of Schwann cell plasma membrane warpped around the axons.

Individual leaflets of the **myelin sheath** are similar to other plasma membranes in being composed of lipids and proteins, but the molecular species and proportions are different. Lipids comprise about 75 percent of myelin and include cholesterol, phospholipids, and galactocerebrosides. Myelin contains less protein than other cell membranes (25 percent), and much of the protein is a distinctive **myelin basic protein,** which becomes a strong autoantigen in certain cases of allergic encephalitis.

A myelin sheath begins at the base of the axon hillock and ends just before the synapse. There are regular interruptions along the course of the axon known as **nodes of Ranvier,** where one Schwann cell ends and another takes over. Nodes are responsible for saltatory electrical conduction (see below).

Neurons communicate with other neurons and other cells at **synapses** (Gr. *synapsis,* point of contact). Action potentials arriving at the synapse trigger the release of synaptic vesicles containing neurotransmitters, and neurotransmitters diffuse across the synaptic cleft to depolarize the postsynaptic membrane.

Axons are also necessary to maintain the integrity of innervated structures. For instance, motor axons are necessary for structural and functional integrity of muscles, supplying a so-called **trophic** influence. Striated muscles atrophy in the absence of innervation—much more atrophy than can be accounted for by simple disuse. The mediator of neurotrophic effects remains obscure.

MEMBRANE POTENTIAL

The **membrane potential** is a voltage difference of about 60 millivolts (negative inside) across the plasma membrane of neurons. Any voltage difference depends on electric charge distribution and flux; Na^+, K^+, Cl^-, and Ca^{2+} carry most electric charge in neurons. Neurons distribute these ions across the plasma membrane using (1) metabolic energy, (2) the Na^+-K^+ATPase carrier, and (3) special voltage-gated protein channels in the membrane; Na^+ and K^+ are the most important ions for the membrane potential.

Neurons (and all cells for that matter) continuously pump Na^+ ions out of the cytosol and K^+ in with a membrane carrier protein, **Na^+-K^+ATPase**. Pumping causes Na^+ to be about ten times lower inside than outside the cell, and K^+ is about ten times higher inside than outside.

If the membrane were permeable only to K^+ ions, the membrane potential

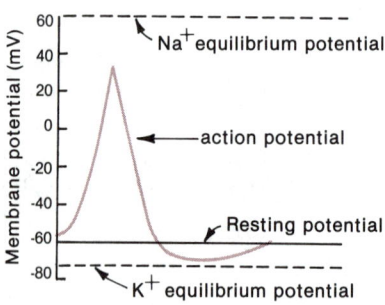

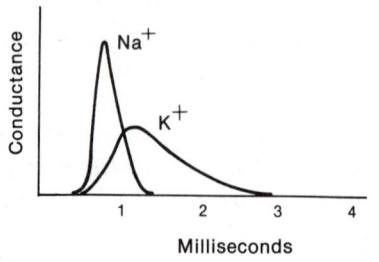

FIG. 7.2. Sodium and potassium conductance generating the action potential. As the Na^+ conductance increases, the membrane potential depolarizes and approaches the Na^+ equilibrium potential. The membrane is repolarized by decreasing Na^+ conductance and increasing K^+ conductance, driving the membrane potential towards the K^+ equilibrium potential.

would be determined only by potassium and would equal the K^+ equilibrium potential of -75 millivolts. If, however, the membrane were permeable only to Na^+ ions, the membrane potential would equal the Na^+ equilibrium potential, which is $+60$ millivolts. In actuality, the membrane is much more permeable to K^+ than to Na^+ ions, so the actual flux of K^+ greatly exceeds that of Na^+, and the potassium equilibrium potential largely determines the resting membrane potential (Fig. 7.2). This is why the potential is -60 millivolts inside.

ACTION POTENTIAL

Hodgkin and Huxley discovered that the **action potential** is an electrical impulse (signal) produced by a brief flow of Na^+ and K^+ ions as a result of transient changes in membrane permeability. The process can start only after Na^+-K^+ATPase builds up a charge across the membrane by decreasing Na^+ and increasing K^+ inside the cell. Once the concentrations are established, ions can flow under the influence of their electrochemical gradient (voltage gradient and concentration gradient). A flow of ions is the same as a flow of electric current, and a transient flow of electric current (change in potential difference) is the same as an electrical impulse. The action potential is a transient flow of electric current involving (1) protein channels that are selective for specific ions and (2) voltage gating of the channels.

Voltage-gated channels control membrane permeability to Na^+ and K^+ during the action potential. The gates are opened at certain voltages, and ions flow through the channels, but the gates close at other voltages, and ion flow stops. These channels are protein complexes that completely extend across the membrane. The Na^+ channel is a large multisubunit complex with a molecular weight of 230,000, and it has allosteric properties that are voltage regulated (gated).

An action potential begins with a momentary depolarization of a small section of neuronal membrane, such as a patch of postsynaptic dendritic membrane. If depolarization is above a threshold voltage (-40 millivolts), adjacent voltage-gated Na^+ channels open, and Na^+ flows into the cell, causing further depolarization by shifting the membrane potential towards the Na^+ equilibrium potential. Once triggered, the action potential travels "all or none" without distortion or attenuation over the entire cell membrane.

Repolarization is also a voltage-dependent process that occurs in the wake of depolarization and involves closing and inactivating Na^+ gates and opening K^+ gates. Potassium ions flow through the open K^+ channels and bring the membrane potential back towards the K^+ equilibrium potential, thus repolarizing the membrane. Sodium channels recover from inactivation, and the membrane is ready to transmit another impulse in less than a millisecond (Fig. 7.2).

Myelination increases the velocity of the action potential by induing saltatory conduction. A Schwann cell covers a segment of axon with myelin for a distance of about 1 millimeter, preventing current flow over the covered neuronal membrane. Between adjacent Schwann cells, a small area of axon is bare at the nodes of Ranvier, and Na^+ channels are concentrated in these areas (as opposed to nonmyelinated fibers, which have a homogeneous distribution of channels over the entire axon). Myelination makes the axon an excellent cable with low capacitance and high resistance to current leakage at the covered segments. The voltage drop is rapidly transferred from one node to the next by **saltatory conduction** (L. *saltatis*, to jump). In myelinated fibers, action potentials travel faster and require much less energy since the ion flow is confined to small nodal areas.

SYNAPSES

Nerves communicate chemically and electrically at **synapses.** A synapse is the terminal enlargement of an axon where one nerve makes contact with another nerve or with an effector organ. The most common type of synapse is an axodendritic synapse between an axon of one neuron and a dendrite of another neuron. Synapses may also be axosomatic (on the cell body) or axoaxonic. A synaptic cleft of 20 to 500 angstroms separates the presynaptic from the postsynaptic membrane. The presynaptic axonal cytoplasm contains synaptic vesicles full of neurotransmitters.

Synaptic transmission involves (1) receptor proteins linked to ion channels and (2) calcium-mediated exocytosis. Arrival of the action potential at the synapse opens Ca^{2+} channels in the presynaptic membrane, and a transient calcium current causes synaptic vesicles to approach and fuse with the presynaptic membrane, dumping neurotransmitters into the synaptic cleft. Neurotransmitters diffuse across the synaptic cleft and bind to receptor proteins on the postsynaptic membrane, opening Na^+ channels, depolarizing the membrane, and initiating a new action potential. The best-characterized receptor is the acetylcholine receptor of the neuromuscular junction (a pentameric glycoprotein of molecular weight 250,000).

Excitatory neurotransmitters that can depolarize the postsynaptic membrane include acetylcholine, dopamine, norepinephrine, and epinephrine. **Inhibitory neurotransmitters** hyperpolarize the membrane by admitting Cl^- ions. The most common inhibitory neurotransmitter is γ-aminobutyric acid, also called GABA.

Only one axon leaves the neuronal cell body at the axon hillock, but numerous terminal branches of an axon permit one neuron potentially to make thousands of synapses. A neuron, an axon, and the muscle fibers it innervates comprise a **motor unit.** Branching of axons determines the fine motor quality of a motor unit. For example, one neuron branches to innervate over 2,000 muscle fibers in the gastrocnemius, giving an innervation ratio of over 1:2,000. However, the extraoccular motor unit of the eye is finely controlled with an innervation ratio of about 1:10.

PERIPHERAL NERVES

Most peripheral nerves are clusters of mixed, afferent and efferent, sensory and motor axons bound together by three distinct connective tissue layers, the **endoneurium, perineurium,** and **epineurium.** The innermost connective tissue layer is the endoneurium, which surrounds individual axons and their Schwann cells. Next is the perineurium, which collects individual axons into discrete bundles called fascicles. Finally, the epineurium binds the fascicles together and surrounds the entire nerve structure with a circumferential fibrous sheath. Vessels and lymphatics run within the endoneurium, providing nutrition to the Schwann cells.

Peripheral motor axons within the endoneurium may be over a meter in length, as in the case of anterior horn cell axons to the intrinsic muscles of the foot. Even more impressive is a sensory bipolar dorsal root ganglion cell with one axonal process in the toe and another in the caudate nucleus of the cervical cord.

Nerve conduction velocity is directly proportional to axonal diameter, and the most widely accepted classification of peripheral nerve fibers uses conduction velocity and axon diameter to divide axons into A, B, and C fibers (Table 7.1). **A fibers** are the largest and fastest. They include myelinated

TABLE 7.1. *Peripheral nerve classification*

Class	Axon diameter (micrometers)	Speed (meters per second)	Innervation
Motor axons			
Aα	12–20	65–120	Fast-twitch extrafusal fibers
Aβ	7–14	40–80	Slow-twitch extrafusal fibers
Aγ	2–10	10–50	Muscle spindle intrafusal fibers
B	1–5	4–25	Presynaptic autonomics
C	0.2–0.5	0.2–2.0	Postsynaptic autonomics
Sensory axons			
Ia	12–22	65–130	Muscle spindles
Ib	12–22	65–130	Golgi tendon organs
II	5–15	20–90	Pressure, touch receptors
III	2–10	12–45	Temperature, pain
IV	0.2–1.5	0.2–2.0	Pain, visceral

Adapted with permission from Mathers (1985).

somatic afferent and efferent axons. **B fibers** are myelinated preganglionic fibers of the autonomic nervous system, and **C fibers** are nonmyelinated sensory and autonomic fibers. Most pain fibers are small, slow-conduction class C fibers.

Reflex Arc

In a reflex arc, propagation of action potentials into the spinal cord over afferent fibers results in the discharge of action potentials over efferent fibers, which produces a simple action. A typical reflex arc occupies a spinal segment and requires either four or five neural elements: (1) a peripheral receptor, (2) an afferent sensory neuron, (3) internuncial neurons (except for the stretch reflex), (4) an efferent motor neuron, and (5) a terminal effector (muscle, gland).

Accidentally touching a hot object illustrates a reflex arc. Finger pain receptors send action potentials up the afferent axons, through the dorsal root, and into the cord to activate interneurons, which synapse with motor neurons. Motor neuron action potentials exit the cord via the ventral root to activate agonist muscles, and the hand is pulled away. Simultaneously, interneurons activate inhibitory neurons to the antagonistic muscles and send impulses up the cord to sensory areas of the brain where the pain is "felt." All this takes place in a few milliseconds.

Proprioceptors

Proprioceptors monitor the spatial position of the body by innervating tendons, ligaments, and joint capsules. They include **Pacinian corpuscles, Golgi tendon organs,** and **muscle spindles.** Pacinian corpuscles are concentrated around joint capsules, where they monitor pressure, Golgi tendon organs are located within a tendon near the muscle–tendon junction, where they monitor stretch. Muscles spindles, embedded within striated muscles, detect, respond to, and control changes in muscle length, regulating resting tone.

Spindles consist of six to 14 specialized muscle fibers encased in a connective tissue shell. These muscle fibers do not contribute to the force of contraction; they specifically monitor stretch and set the resting tension of muscle. **Intrafusal muscle fibers** are those inside the spindle, and **extrafusal fibers** are the bulk, force-generating muscle fibers outside the spindle. Motor axons of the class $A\gamma$ (γ efferents) innervate the intrafusal muscle fibers, and Ia sensory axons leave the spindle, conveying stretch information to the CNS.

Muscle Tone

A normal, relaxed muscle still has some residual tension and is not completely flabby. Passive stretching of the muscle by joint motion encounters a certain small amount of resistance. These two characteristics reflect muscle tone, namely, (1) slight resting tension and (2) involuntary resistance to mechanical stretch.

The **stretch reflex** sets the tension in muscles and determines normal resting muscle tone. The stretch reflex is unique because its afferent and efferent neurons make direct synapse without the use of intervening interneurons. Stretching a muscle activates the muscle spindles, and action potentials travel up the Ia fiber to the anterior horn gray matter, where they synapse with α motor neurons. Efferent Aα motor axons conduct action potentials from the cord to the motor end plates, activating the muscle and increasing the tension. A muscle's stretch reflex is tested by tapping the tendon to give a slight, quick stretch (Fig. 7.3) and observing the quality of the involuntary contraction.

The γ efferents cause contraction of intrafusal muscle fibers of the spindle, and a constant feedback between intrafusal and extrafusal fibers sets the tone of the muscle. Motor centers in the brain control the firing of the γ efferents. By increasing the tone of intrafusal fibers, the spindle becomes more sensitive to stretch and increases reflex tone of the muscle. Decreasing the tone of the intrafusal fibers relaxes the tone of the entire muscle.

Balanced input of central descending tracts from the brain and brainstem

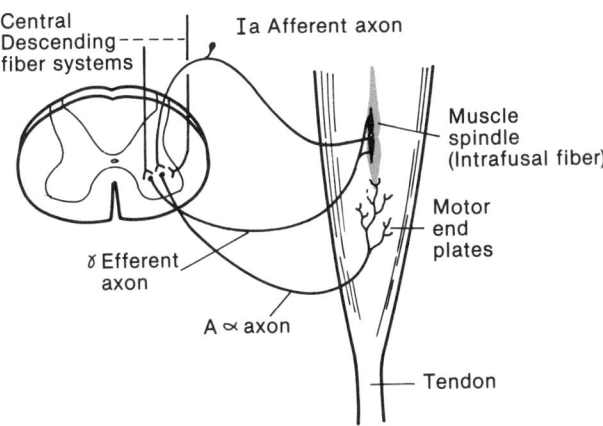

FIG. 7.3. The stretch reflex. Stretching the spindle activates the Ia afferent neuron, which synapses on the α motor neuron, producing efferent action potentials that cause the muscle to contract, relaxing the stretch on the spindle. Adapted with permission from Truex and Carpenter (1969).

control the stretch reflex. Injury to these central tracts results in uncontrolled stretch reflex activity and **spastic** paralysis. Muscles are firm and resist passive stretching; quick tendon stretch causes clonus. On the other hand, injury to anterior horn cells, ventral roots, or peripheral nerve motor fibers abolishes reflex tone and causes **flaccid** paralysis. A flaccid muscle is severely atrophic with marked fatty and connective tissue replacement of muscle fibers.

Mechanical Nerve Injury

Seddon classified peripheral nerve injuries into three categories based on clinicopathological response: neuropraxia, axonotmesis, and neurotmesis. **Neuropraxia** is the least involved injury and occurs following an event such as compression tourniquet palsy. Compression injuries Schwann cells and disrupts the myelin sheath, but axons and connective tissue tubes remain intact. Nerve conduction is abnormal across the injured segment but normal proximal and distal. Recovery occurs in a matter of days to weeks (typically 10 to 14 days), and nerve function returns to normal.

Axonotmesis is a physical separation of the distal part of an axon from the cell body but with preservation of connective tissue tubes (endoneurium, perineurium, and epineurium). It is usually caused by a crush injury or by chronic compression such as a carpal tunnel syndrome. Axonal transport is disrupted, axons and myelin distal to the injury undergo **Wallerian degeneration,** and phagocytosis eventually removes the residual cellular debris. Neuronal cell bodies also change by enlarging and undergoing a process of **chromatolysis,** in which the cell body ultrastructure and metabolism change from a steady state to a regeneration state. Nerve conduction is abnormal from the site of injury to the end of the nerve. Recovery begins as axon sprouts grow down the endoneural tubes. The process proceeds at a rate of about 1 millimeter per day or 2.5 centimeters per month. With the return of normal myelination, nerve function returns to normal.

Neurotmesis is the disruption of anatomic integrity of the nerve such as occurs with a cut or following violent traction. Axons, myelin, and the connective tissue layers are all torn. Both Wallerian degeneration and chromatolysis occur, and without surgical repair, a random growth of axon sprouts, Schwann cells, and fibroblasts results in a **neuroma.** Following surgical repair some axon sprouts grow across the injury and eventually reestablish peripheral contact, but recovery is incomplete, and functional outcome is usually less than normal.

Neuropathies

Mononeuropathies are focal nerve injuries caused by mechanical pressure, entrapment, or ischemia (Table 7.2). Polyneuropathies are diffuse nerve dis-

TABLE 7.2. *Sensory and motor events after tourniquet ischemia*

	After inflation		After deflation
10 minutes	Fingertip numbness	30 seconds	Return of sensation
15 minutes	Loss of light touch, pressure	4–6 minutes	Intense paresthesias
30 minutes	Loss of pain, temperature	10 minutes	Return of muscle power
40 minutes	Complete paralysis		

Extrapolated from Lewis et al. (1931).

orders resulting in muscle atrophy, sensory changes, and reflex loss. Paresthesias, pain, and numbness are common, usually affecting feet and legs before hands and arms and distal before proximal areas. Neuropathies may be caused by demyelination or axon degeneration (Table 7.3). They may be either genetic or acquired, and they may present as acute, subacute, or chronic illness. Nerve conduction velocity is delayed, and the electromyographic pattern is one of irritability with prolonged insertional activity and fibrillations.

MUSCLE STRUCTURE

The human body contains more than 430 muscles classified into three groups: cardiac, smooth, and striated. Cardiac muscle occurs in ventricular and atrial walls. Smooth muscle (nonstriated, involuntary) occurs in gut and blood vessel walls and gives prolonged, tonic contractions. Striated muscle, also called skeletal muscle, constitutes 40% of the total body weight and powers the axial and appendicular skeleton, producing voluntary body motions by converting the chemical energy of ATP into the mechanical energy of contraction.

A muscle such as the biceps brachii is composed of many multinucleated cells called **muscle fibers.** Muscle fibers arise from the fusion of hundreds of myoblasts, and each fiber retains all the myoblast nuclei, some fibers having over 100 flattened nuclei lying peripherally along the cell membrane (the sarcolemma). The muscle fibers are arranged into bundles called **fasciculi.**

TABLE 7.3. *Some common neuropathies caused by demyelination and axon degeneration*

Category	Demyelination	Axon degeneration
Genetic	Refsum's disease	Friedreich's ataxia
Toxic	Diphtheria	Heavy metals, vincristine
Metabolic	Hypothroidism	Diabetes, uremia
Inflammatory	Guillain–Barré	Collagen vascular disease

Skeletal muscles have an origin and an insertion relative to the joint of motion. The origin is usually a proximal, fixed attachment site; the insertion is a distal attachment site to the bone that is moved.

Connective tissue runs throughout the muscle from end to end, blending in with the tendons of origin and insertion. The **endomysium** fills the space between muscle fibers. The **perimysium** surrounds up to 150 fibers, creating the fascicles and holding them together. The **epimysium** surrounds and binds the entire muscle and is continuous with the epitenon (the tendon sheath).

Nerve supply and the major vessels of a muscle usually enter the deep surface near the origin at a fairly constant position called the **neurovascular hiatus.** Vessels branch within the perimysium and divide into a network of capillaries that run parallel to the fibrils in the endomysium. Lymphatic capillaries also are present in the epimysium and perimysium but apparently do not penetrate to the endomysium. In rhythmic exercises such as cycling, running, or swimming, blood flows mostly during the relaxation phase. Static or isometric exercises decrease or in some cases can actually stop the flow of blood, necessitating anaerobic metabolism. Endurance training increases the capillary density, and capillaries can be up to 40% greater in elite marathon runners than in untrained controls.

The nerve to a muscle contains both motor and sensory fibers, including large myelinated efferents of the anterior horn neurons (α efferents), γ efferents to the muscle spindles, and nonmyelinated autonomic efferents to the vascular smooth muscle. The α efferents branch to innervate a variable number of individual muscle fibers at the motor end plate (also called the neuromuscular junction). Action potentials traveling down the α efferent axon cause release of acetylcholine at the motor end plate, depolarizing the postsynaptic membrane. Depolarization spreads over the entire fiber surface and into the transverse (T) tubules, stimulating the sacroplasmic reticulum to release calcium, which triggers contraction of the myofibrils (more on T-tubules below).

MUSCLE ULTRASTRUCTURE

Composition analysis shows that skeletal muscle is 75 percent water, 20 percent protein, and 5 percent inorganic salts including high-energy phosphates, lactate, and carbohydrates. The most abundant proteins are myosin (52%), actin (23%), tropomyosin (15%), and myoglobin (1%). **Myoglobin,** a heme protein, serves as a reserve supplier of oxygen and facilitates oxygen movement within muscle fibers. Myoglobin along with the cytochromes gives muscle its characteristic red color.

Muscle cytoplasm (sarcoplasm) contains myofibrils, glycogen granules, ATP, phosphocreatine, glycolytic enzymes, and mitochondria. Myofibrils are the contractile elements of muscle and are approximately 1 micrometer (1/1000 millimeter) in diameter. They occupy most of the sarcoplasm and extend the entire length of muscle fibers. Under the light microscope, myofibrils have a striated appearance created by alternating light and dark bands. These bands were named according to their influence on polarized light. The **dark A bands** strongly rotate the plane of polarized light away from the eye (A stands for anisotropic), and the **light I bands** have little influence on the polarized light (I for isotropic). I bands are bisected by **Z lines** (Zwischenscheibe), and the A bands are bisected by Hensen's zones or **H zones.** The distance between two Z lines establishes a repeating unit called the **sarcomere** (Fig. 7.4).

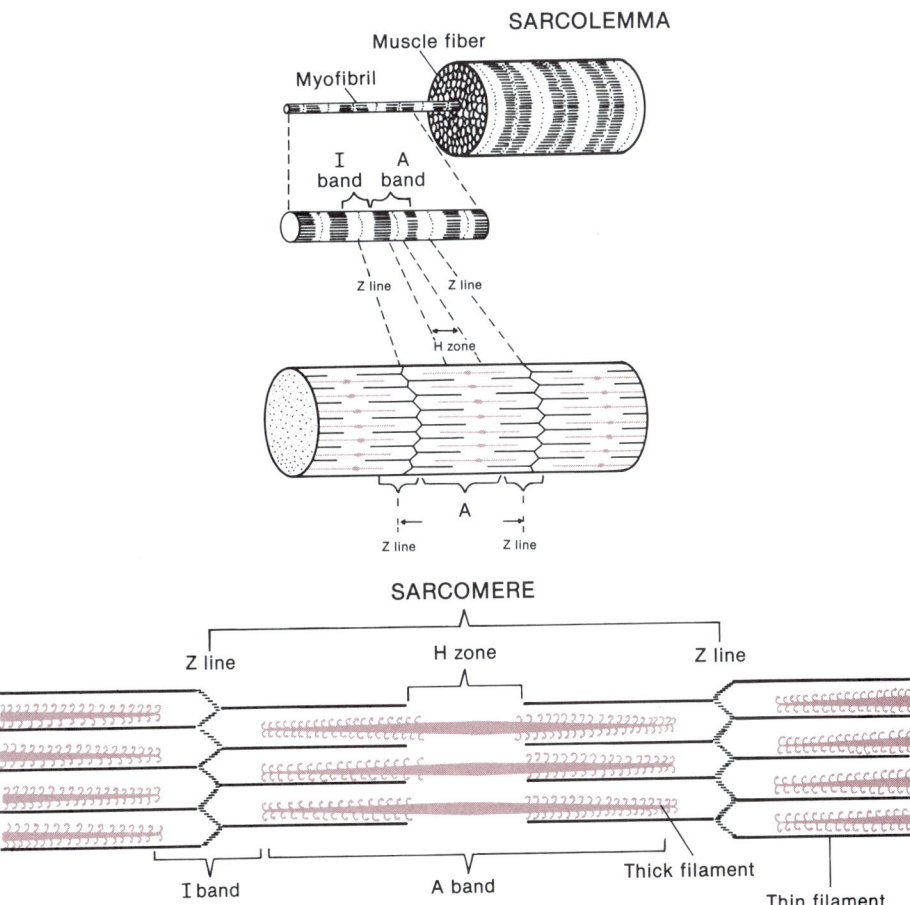

FIG. 7.4. Structure of the sarcomere. Myosin thick filaments form cross bridges with actin thin filaments. Conformational changes in myosin heads after binding to actin thin filament produce contraction. Adapted with permission from Spence and Mason (1979).

Sarcomere Structure and Function

Electron micrographs and biochemical studies show that the sarcomere is made of two sets of parallel and partly overlapping protein filaments. **Thick filaments** extend from one end of the A band to the other and consist primarily of myosin. **Thin filaments** are attached to the Z line and extend across the I bands and partially into the A bands. Thin filaments contain actin, tropomyosin, and troponin. Z lines contain a binding protein called α-actinin, and M zones contain a special protein called M-protein. Myosin thick filaments are held together at the M zones.

H. E. Huxley, A. Huxley, and co-workers proposed the **sliding filament model** to account for muscle contraction. It was known that a sarcomere shortens as muscle contracts, but the lengths of the thick and thin filaments do not change. Only the I band decreases in length; the A band remains unchanged as the distance between Z lines decreases. Huxley proposed that the thick and thin filaments slide past each other during contraction, and the force of contraction is generated by the process that moves the filaments. Subsequent studies have confirmed that mechanical tension varies according to the amount of overlap between the thick and thin filaments. Very-high-magnification electron micrographs show that the thick myosin filaments have tiny side arms that form cross bridges with the thin actin filaments, and movement occurs through an oar-like or ratchet-like interaction between the two filaments (Fig. 7.4).

Both myosin thick filaments and actin thin filaments are assembled from subunits. Myosin subunits have a molecular weight of about 500,000 and are rod-like molecules composed of two globular heads attached to a long filamentous tail. Myosin molecules associate with each other by their tails, and each thick filament contains about 500 myosin tails packed together with the myosin heads sticking out. Actin has globular subunits (G-actin) of molecular weight 41,800 that polymerize into a helical thin filament (F-actin). Each G-actin subunit has a specific myosin-binding site for a myosin head. F-actin binds two other proteins involved in triggering muscle contraction, troponin and tropomyosin.

During contraction, myosin heads bind to actin filaments at the specific binding sites, and the myosin heads "walk" along the actin filaments by making and breaking bonds to successive binding sites. The power stroke comes when a myosin head bound to an actin filament undergoes a conformational change causing the head to pull against the thin filament. Acting in concert, hundreds of myosin heads pull the actin filaments towards the H zone and shorten the sarcomere. Energy for the conformational change and muscle contraction comes from hydrolysis of ATP to ADP; the efficiency is high, in the range of 30 to 50% (automobile engines run at about 10%).

Tropomyosin and troponin are two accessory proteins that control contraction by preventing the interaction of myosin with actin. Tropomyosin binds to sites in the helical groove along the entire length of an actin thin filament and blocks the myosin-binding sites. Troponin binds to both tropomyosin and actin. Troponin has a high affinity for calcium ion. When calcium binds to troponin, the tropomyosin–troponin complex loses its affinity for actin, exposing myosin-binding sites, and cross bridging between myosin and actin can take place.

EXCITATION–CONTRACTION COUPLING

The process by which depolarization of muscle fibers causes myofibril contraction is referred to as **excitation–contraction coupling,** and it requires two membrane specializations: (1) transverse or T-tubules and (2) sarcoplasmic reticulum (Fig. 7.5). T-tubules are plasma membrane specializations that extend into the sarcoplasm and lie adjacent to the terminal cisternae of sarcoplasmic reticulum. Sarcoplasmic reticulum is a network of flattened vesicles derived from the endoplasmic reticulum. The sarcoplasmic reticulum sequesters high concentrations of calcium and surrounds each myofibril like an irregular net stocking (Fig. 7.5).

A muscle action potential spreads over the sarcolemma and down T-

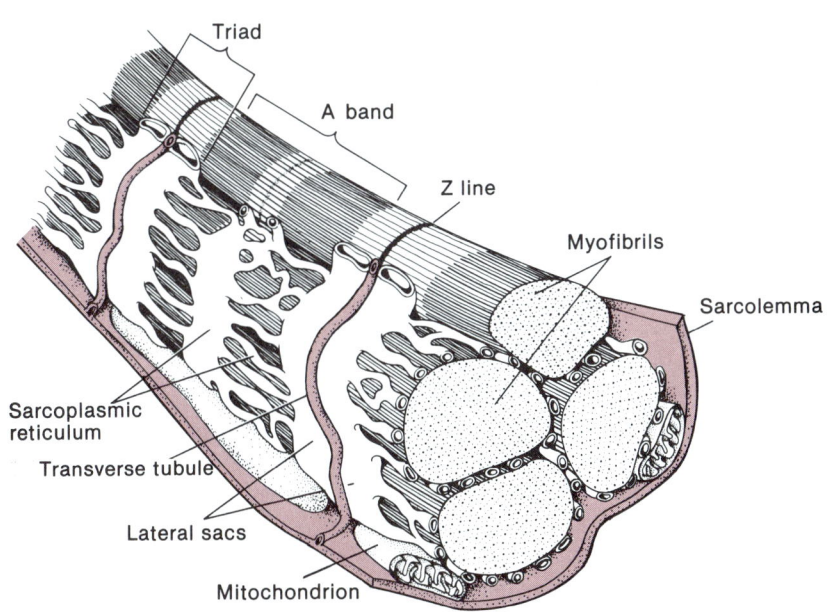

FIG. 7.5. Transverse tubules and sarcoplasmic reticulum of a muscle fiber. The T-tubule is an extension of the plasma membrane deep into the cell, where it forms a triad (T-tubule plus two sarcoplasmic reticulum lateral sacs). This system speeds and synchronizes the release of calcium. Adapted with permission from Spence and Mason (1979).

TABLE 7.4. *Events of muscle contraction and relaxation*

1. Acetylcholine released at NMJ
2. Muscle membrane and T-tubules depolarized
3. Ca^{2+} released from sarcoplasmic reticulum
4. Ca^{2+} binds to tropomyosin–troponin complex
5. Steric block of actin removed
6. Myosin heads bind to actin
7. ATP powers myosin conformational changes
8. Actin filaments pulled against thick filament
9. Sarcomere shortens by filament overlapping
10. Ca^{2+} pumped out of cytoplasm
11. Tropomyosin–troponin blocks myosin binding
12. Muscle relaxes

tubules, activating the sarcoplasmic reticulum and causing release of large amounts of calcium into the vicinity of the myofibrils. It is the sudden rise in free calcium concentration that initiates myofibril contraction by removing the tropomyosin–troponin block. Myosin heads, which project from thick filaments, now cross bridge with actin filaments. In an ATP-driven cycle, myosin heads undergo a conformational change that pulls one filament against the next. Repeated cycles of attachment, pulling, and detachment take place to shorten the sarcomere, contract the muscle, and move the body part.

Contraction is terminated and fibers relax when calcium is sequestered back in the sarcoplasmic reticulum cisterns. ATP-driven calcium carrier proteins pump calcium out of the cytoplasm and back into the sarcoplasmic reticulum, where proteins called calsequestrin, each of which has 40 calcium sites, bind the free calcium. Tropomyosin–troponin then resumes the steric block of myosin-binding sites, cross bridges can no longer form, and the muscle relaxes (Table 7.4).

ENERGY METABOLISM

ATP is the direct energy source for muscular contraction. Myosin heads have an ATPase enzyme that splits ATP to ADP and inorganic phosphate, using the released free energy to power a conformational change that pulls the thin filament past the thick filament. If a cell runs out of ATP, contraction cannot occur. Furthermore, the amount of ATP normally present within a muscle can sustain contraction for only a fraction of a second. To prevent ATP depletion, muscle fibers use three enzyme systems to maintain the ATP charge (1) the creatine phosphate reservoir, (2) glycolysis, and (3) mitochondrial oxidative phosphorylation.

Phosphocreatine, a high-energy phosphate compound found in muscle, acts as a rapid buffer for the resynthesis of ATP. **Creatine kinase** is an enzyme attached to the M zone of myosin, and it catalyzes the transfer of a phosphoryl group from phosphocreatine to ADP to form ATP according to the reaction:

$$\text{Phosphocreatine} + \text{ADP} \underset{}{\overset{\text{Creatine kinase}}{\rightleftarrows}} \text{ATP} + \text{Creatine}$$

Creatine is recharged to phosphocreatine with ATP derived either from glycolysis or mitochondrial oxidative phosphorylation. The relative contribution of glycolysis or oxidative phosphorylation depends on the muscle fiber type. As we shall see, red muscle uses mostly aerobic, oxidative metabolism, and white muscle uses anaerobic glycolysis.

MUSCLE FIBER TYPE

As mentioned in the section on synapses (above), a motor unit is an α efferent neuron, its axon, plus all the muscle fibers on which it synapses. The number of muscle fibers innervated varies, being about ten in the extraoccular muscles and more than 1,000 in the leg antigravity muscles. Motor units also vary functionally and biochemically. Based on contraction velocity and metabolic characteristics, muscle fibers are classified as either **type I (slow twitch, oxidative)** or **type II (fast twitch, glycolytic)**. The neuron that innervates the fibers actually determines whether the fibers will be slow or fast twitch, presumably as part of the trophic influence. In most muscles, slow and fast twitch fibers are intermingled in a more or less random mosaic, but all fibers of the same motor unit have identical functional and biochemical characteristics.

Type I fibers have a relatively slow contraction time (100–120 milliseconds) and are resistant to fatigue. They contain numerous mitochondria high in Krebs cycle and oxidative phosphorylation enzymes. Metabolism is mainly aerobic, oxidative, and the capillary density is high around the fibers. Type I fibers are best adapted for prolonged activity of low intensity, and the motor neuron tends to fire more on a tonic than on a phasic basis. Type I fibers are referred to as "red" fibers in the older literature.

Type II fibers are fast twitch (40 milliseconds), rapidly fatiguing fibers best adapted to short-duration, high-intensity activity. They contain numerous intracellular glycogen granules and have high concentrations of both glycogen and glucose metabolism enzymes such as phosphorylase and lactic dehydrogenase. Metabolism is mostly anaerobic, and mitochondria are sparse, but type II fibers have a highly developed transverse tubular system, which facilitates rapid contraction. Type II fibers fatigue rapidly because they rely mainly on stored intracellular glycogen for energy. The motor neuron tends to fire on a phasic basis.

Type II fibers are further subdivided into IIa and IIb fibers. **Type IIa**

TABLE 7.5. *Functional and biochemical characteristics of muscle fiber types*

Muscle type	Contraction speed	Color	Oxidative capacity	Glycolytic capacity
I	Slow	Red	High	Low
IIa	Fast	Red	High	High
IIb	Fast	White	Low	High

have high glycolytic and high oxidative enzymes; **type IIb** fibers are predominantly glycolytic in metabolism. Both IIa and IIb fibers have the same fast twitch characteristic. Type II fibers are occasionally referred to as "white" fibers (Table 7.5).

Histological differences in the composition of skeletal muscle account for some of the diversity in human performance capacity. For instance, the average person has about 50 percent type I, 25 percent IIa, and 25 percent IIb fibers in his calf muscles. However, elite marathon runners have more than 90 percent type I fibers, and world-class sprinters have predominantly fast-twitch type IIa fibers. Many other factors are important for athletic performance, such as synchronicity of motor unit activation, cardiorespiratory capacity, and psychological preparation, but muscle fiber composition has a major influence on basic muscular ability. Most evidence suggests that the percentage distribution of muscle fiber types is genetically determined and not substantially changed by training.

CONTRACTILE PROPERTIES

Muscle fiber type is not the only factor to determine contraction velocity. In general, velocity is proportional to the number of sarcomeres acting in series, and muscle force is proportional to the number of sarcomeres acting in parallel. This means that elongated muscles with long fibers have more sarcomeres in series, and they have greater contraction velocities than short muscles. Similarly, short thick muscles have more fibers in parallel, which means more sarcomeres in parallel, and they generate greater muscle force. Thus, long muscles such as the sartorius and brachialis are designed for high-velocity output. The vastus, deltoid, and soleus muscles, with pennated architecture and short fiber length, are designed for high force output.

Another important property is the **length–tension** relationship. This relates the initial muscle length to the maximum tension that it develops during contraction. The **Blix curve** is an expression of the length–tension relationship, plotting isometric muscle tension versus the ratio of contraction length to initial length. This curve graphically demonstrates that the tension

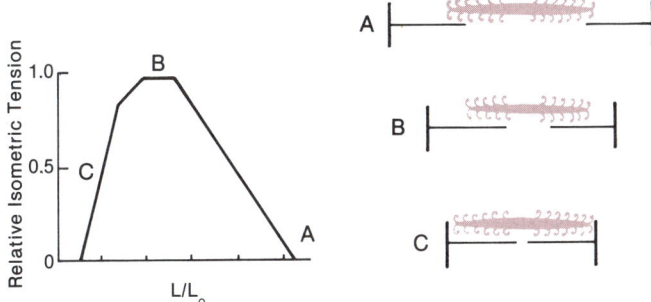

FIG. 7.6. The length–tension relationship. Maximum tension develops at the resting length (point B). Isometric tension sharply declines at lengths greater or less than the resting length. Modified from Gordon et al. (1966).

(force) developed is proportional to the number of available interaction sites between myosin and actin filaments. Thus, the greatest tension develops at the resting length, where the maximum possible cross bridges can form. At shorter lengths or longer lengths, fewer cross bridge sites are available, and less tension develops (Fig. 7.6). This relationship has important implications for surgeons who must reattach muscles at the correct length after trauma or tendon transfer.

EFFECTS OF TRAINING

Various training techniques focus on muscular strength, endurance, power, flexibility, or combinations, but all techniques rely on muscle contraction, of which there are three fundamental varieties: isometric, isotonic, and eccentric. **Isometric contractions** occur when tension develops but the muscle is held at a fixed length; i.e., no work is done. **Isotonic contractions** occur when a muscle shortens under a constant load, and **eccentric contractions** involve maintaining tension while the muscle is being stretched, such as decelerating the leg during gait. In athletic training, it is difficult to establish true isometric conditions because there is always some joint motion, and it is almost impossible to establish an isotonic condition since the tension will vary throughout the range of motion. Nonetheless, these terms are used to describe exercise techniques.

Isometric exercise involves the contraction of muscles around a joint with little or no joint motion. These exercises can be of some benefit when joint motion is contraindicated, but isometrics have little effect on type II fibers.

Isotonic exercise involves contraction of a muscle with joint motion such that the tension remains relatively constant as the muscle shortens. Isotonic exercise methods can utilize constant, variable, speed, plyometric, and eccentric resistance. With constant resistance, such as using barbells, the load remains constant, but the difficulty overcoming resistance varies with joint angle and is usually most difficult at the beginning of motion. Variable-resistance exercise uses specially designed machines such as the Nautilus® with a changing fulcrum and lever arm to increase the load throughout the range of motion. Speed loading involves moving the resistance as fast as possible; it is inferior to progressive resistive exercise for gaining strength but is quite valuable for endurance. Plyometric resistance involves a sudden loading of muscles and forced stretching before maximum contraction, an example of which is jumping from a block and then rebounding back up on the block. In eccentric isotonic resistance, the muscle is contracted during lengthening. Eccentrics produce more muscle soreness (inflammation) and seem to offer no superiority to other techniques.

Isokinetic exercise controls the rate of muscle shortening by resisting the exerted muscle force with a slightly smaller force. With special dynamometer exercise machines such as a Cybex II®, the angular velocity of joint motion is preset, and the machine accommodates resistance to whatever torque the patient applies. Isokinetics are very popular because they allow rehabilitation and training of injured joints with less risk of further injury.

Athletic training demonstrates fiber specificity. High-intensity, low-repetition exercise (weight lifting) causes hypertrophy of fast-twitch type II fibers with only minor changes in type I fibers. Power lifters have type II fibers that are 45% larger than those of age-matched controls; the hypertrophy is caused by increased actin and myosin filaments and increased glycogen content. On the other hand, high-repetition, lower-intensity exercise (distance running) specifically recruits slow-twitch type I fibers, increasing the number and size of mitochondria and improving oxidative capacity. This specificity of training is one of the reasons for poor carryover of training from one sport to another (i.e., a highly trained swimmer is not necessarily a good runner).

Muscle Atrophy

Muscles atrophy from decreased activity (disuse), immobilization, starvation, or denervation. Loss of innervation removes the trophic influence, and muscle degenerates with replacement by fat and fibrous tissue. Disuse or immobilization causes loss of bulk, but fibers retain their external integrity. In disuse atrophy, fibers decrease in size by loss of both contractile protein and sarcoplasmic reticulum. Immobilization produces a faster atrophy of type I than of type II fibers.

GLOSSARY

A bands The dark, anisotrophic muscle bands corresponding to the myosin thick filaments.

Action potential An all-or-none electrical depolarization of a nerve or muscle cell membrane caused by transient changes in membrane permeability to sodium and potassium.
Afferent Towards the central nervous system, as in axons that convey sensory information to the spinal cord.
α motor neuron Large motor neurons in the ventral gray matter of the spinal cord, innervating skeletal muscle extrafusal fibers.
Axon The long process of neurons conducting action potentials away from the cell body.
Axon hillock The pyramidal protrusion of a neuron cell body where the axon arises.
Axonotmesis A physical separation of the distal part of an axon from the cell body but with preservation of the connective tissue tubes, often resulting from a crush injury.
Dendrites Unmyelinated processes projecting from neurons to receive synaptic input from other neurons and conduct impulses toward the cell body.
Eccentric contractions Maintaining tension while a muscle is stretched.
Efferent Away from the central nervous system, as in axons that convey action potentials from the spinal cord to the muscles.
Endomysium The innermost connective tissue that surrounds individual muscle fibers.
Endoneurium The innermost connective tissue layer of a peripheral nerve that collects and fills in the space between individual Schwann cells.
Epimysium Connective tissue layer that surrounds the entire muscle.
Epineurium The outermost connective tissue sheath of a peripheral nerve, binding the nerve fascicles together.
Extrafusal fibers The bulk, force-generating fibers of a muscle, outside the muscle spindles and innervated by the Aα efferent axons.
Fibers Multinucleated muscle cells formed by the fusion of numerous myoblasts.
γ motor neurons Neurons in the ventral gray matter of the spinal cord that supply intrafusal fibers of muscle spindles.
H zone A broad area bisecting the A band, where myosin thick filaments are joined together.
I bands The light, isotropic muscle bands corresponding to the actin thin filaments.
Intrafusal fibers Special muscle fibers inside spindles that monitor stretch; innervated by the γ efferent neurons.
Isokinetic exercise The same rate of muscle shortening throughout the entire range of motion, performed with a dynamometer that controls the resistance at any particular speed to match closely the exerted force.
Isometric contraction Muscle develops tension but does not shorten.
Membrane potential A voltage difference of about 60 millivolts across the plasma membrane, negative inside.

Motor unit A single motor neuron, its axon, and all the muscle fibers it innervates.
Myelin Insulating concentric layers of Schwann cell plasma membrane, containing lipids and protein, wrapped around axons.
Myoglobin A heme protein serving as a reserve supplier of oxygen and facilitating oxygen movement within muscle fibers.
Neuromuscular junction Also called the motor end plate, an acetylcholine-mediated synapse between a motor neuron and muscle fiber.
Neuropraxia A temporary disruption of nerve conduction caused by a minor insult such as tourniquet compression.
Neurotmesis A nerve injury consisting of anatomical disruption of integrity, usually a complete severence.
Node of Ranvier The small unmyelinated gap along an axon between two Schwann cells, where saltatory conduction occurs.
Perimysium Connective tissue layer that surrounds muscle fascicles.
Perineurium Connective tissue layer of a peripheral nerve that surrounds the fascicles.
Phosphocreatine A high-energy phosphoryl ester of creatine able to recharge ADP to ATP in the presence of the enzyme creatine kinase.
Proprioceptors Special sensory organs that monitor the position of the body in space, including Pacinian corpuscles, Golgi tendon organs, and muscle spindles.
Saltatory conduction Electrical conduction along myelinated axons in which the voltage drop is rapidly transferred from one node of Ranvier to another.
Sarcomere The basic contractile unit of myofibrils, extending from one Z line to the next, consisting of the A band between two halves of I bands.
Sarcoplasmic reticulum A series of membrane sacs derived from the endoplasmic reticulum, surrounding myofibrils, and capable of releasing or sequestering Ca^{2+} on demand.
Schwann cells Cells that surround and myelinate peripheral axons.
Synapse Area of communication between a neuron and another cell, consisting of presynaptic membrane containing neurotransmitter, synaptic cleft, and postsynaptic membrane.
Thick filaments Aggregates of myosin tails with myosin heads sticking out from the aggregate shaft.
Thin filaments Polymerized form of G-actin, forming a helical fibrous protein F-actin, attached to the z-line.
Triad Two terminal cisterns of sarcoplasmic reticulum and a transverse tubule.
Type I fibers Slow-twitch, fatigue-resistance fibers that tend to contract tonically, with mainly oxidative aerobic metabolism, high in mitochondria, serving as postural muscles.

Type II fibers Fast-twitch, rapidly-fatiguing fibers that contract phasically using mostly anaerobic, glycolytic metabolism.
Z line A line bisecting the I band, defining the limits of the sarcomere.

BIBLIOGRAPHY

Albers, B., Bray, D., Lewis, J., Raff, M., Roberts, K., and Watson, J. D. (1983): Muscle contraction. *Molecular Biology of the Cell,* pp. 550–560. Garland Publishing, New York.
Catterall, W. A. (1984): The molecular basis of neuronal excitability. *Science,* 223:653–661.
Garrett, W. E., Mumma, M., and Lucaveche, C. L. (1983): Ultrastructural differences in human skeletal muscle fiber types. *Orthop. Clin. North Am.,* 14:413–425.
Gollnick, P. D., and Matoba, H. (1984): The muscle fiber composition of skeletal muscle as a predictor of athletic success. An overview. *Am. J. Sports Med.,* 12:212–217.
Gordon, A. M., Huxley, A. F., and Julian, F. J. (1966): The variation in isometric tension with sarcomere length in vertebrate muscle fibers. *J. Physiol (Lond.),* 184:170–192.
Lewis, T., Pickering, G. W., and Rothschild, P. (1931): Centripetal paralysis arising out of arrested blood flow to the limb, including notes on a form of tingling. *Heart,* 16:1.
Liber, R. L. (1966): Skeletal muscle adaptability. 1: Review of basic properties. *Dev. Med. Child Neurol.,* 28:390–397.
Mathers, L. H. (1985): *The Peripheral Nervous System. Structure, Function, and Clinical Correlations.* Addison-Wesley, Menlo Park, CA.
Seddon, J. H. (1972): *Surgical Disorders of Peripheral Nerves,* pp. 32–36. Livingstone, Edinburgh, London.
Spence, A., and Mason, E. B. (1979): *Human Anatomy and Physiology.* Benjamin/Cummings, New York.
Stryer, L. (1981): *Biochemistry,* pp. 815–829. W. H. Freeman, San Francisco.
Sullivan, J. D., Olna, A. E., Rohan, I., and Schulz, J. (1986): The properties of skeletal muscle. *Orthop. Rev.,* 15:17–31.
Truex, R. C., and Carpenter, M. D. (1969): *Human Neuroanatomy.* Williams & Wilkins, Baltimore.

8
Calcium, Phosphorus, Metabolic Bone Disease

BIOCHEMISTRY OF CALCIUM AND PHOSPHORUS

Calcium is the fifth most abundant element in the earth and also the fifth most abundant element in the human body (Table 8.1). The name calcium comes from the Latin word *calx,* meaning lime, the first source of calcium known to the alchemists. As with all alkaline earth metals, calcium has two valence electrons, and only the 2+ ionic charge is important in its chemistry. Calcium has an atomic number of 20 and an atomic weight of 40; it readily forms phosphates (Table 8.2), carbonates, hydroxides, and sulfates and readily complexes with negatively charged macromolecules such as albumin. The atomic radius of calcium is 0.99 angstroms (99 picometers), making it slightly larger than sodium but smaller than potassium (Table 8.3).

TABLE 8.1. *Elemental composition of the earth and the human body expressed as percentage by mass*

Earth		Human body	
Element	Percentage	Element	Percentage
Oxygen	49.1	Oxygen	64.6
Silicon	26.1	Carbon	18.0
Aluminum	7.5	Hydrogen	10.0
Iron	4.7	Nitrogen	3.1
Calcium	3.4	Calcium	1.9
Sodium	2.6	Phosphorus	1.1
Potassium	2.4	Chlorine	0.40
Magnesium	1.9	Potassium	0.36
Hydrogen	0.88	Sulfur	0.25
Titanium	0.58	Sodium	0.11
Chlorine	0.19	Magnesium	0.03
Carbon	0.09	Iron	0.005

TABLE 8.2. *Physiologically important inorganic calcium compounds*

Highest ←   Acidity   → Lowest Lowest ←   Stability   → Highest

$Ca(HPO_4) \cdot 2H_2O$ Dicalcium phosphate dihydrate	$Ca_4H(PO_4)_3$ Octacalcium phosphate	$Ca_9(PO_4)_6$ (var.) Amorphous calcium phosphate	$Ca_3(PO_4)_2$ Tricalcium phosphate	$Ca_{10}(PO_4)_6(OH)_2$ Hydroxyapatite	$Ca_{10}(PO_4)_6F_2$ Fluorapatite

TABLE 8.3. *Ionic Radii*

Ion	Radius (picometers)
K^+	133
Ca^{2+}	99
Na^+	95
Mg^{2+}	65

Calcium functions in both insoluble and soluble forms. Ninety-nine percent of the body's calcium exists as insoluble, hydroxyapatite-like crystals in the skeleton. The remaining 1 percent is soluble ionic Ca^{2+}, serving as a control and signaling mechanism in many biological activities including muscle contraction, nerve conduction, synaptic transmission, hormone action, gastric secretion, and blood coagulation (Table 8.4). Because the correct Ca^{2+} concentration is critical to so many activities, the normal plasma calcium concentration is one of the most closely guarded values in the blood and varies only between narrow limits. Elaborate metabolic checks and balances maintain the total Ca^{2+} concentration around 2.5 millimolar or 10 milligrams per deciliter, half of which is protein bound. The intracellular concentration of Ca^{2+} is three to four orders of magnitude lower than that of the extracellular fluid and plasma.

Phosphorus is the sixth most abundant element in the body but quite rare in the earth, comprising less than 0.1 percent of the earth's mass. The name phosphorus comes from two Greek words, *phos* meaning light and *phorein* meaning to carry. Alchemists named this element because of the brilliant light emitted from its combustion. Phosphorus is a nonmetallic element with an atomic number of 15 and an atomic weight of 30.9. It is highly flammable and readily forms oxygen-containing phosphates.

TABLE 8.4. *Biological levels where calcium functions*

Organ level	(a) Structural component of bone crystals
	(b) Gastric secretion and pepsin activation
	(c) Cofactor in blood coagulation
Cellular level	(a) Membrane integrity and stability
	(b) Neuromuscular excitation
	(c) Cellular adhesion
Subcellular level	(a) Actomyosin activation
	(b) Axonal transport
	(c) Secretion of cellular products
Enzyme level	(a) Adenylate cyclase activation
	(b) Phosphorylase kinase activation
	(c) Phosphoenolpyruvate carboxykinase regulation

Eighty-five percent of the phosphate exists in the skeleton, and the remaining 15 percent is in the soft tissues and plasma. Cells use phosphorus both as a substrate and as a structural component. As a substrate, high-energy phosphate bonds provide the biological currency of energy, driving chemical reactions at every level of cellular metabolism. As a structural component, phosphorus appears in cellular membranes as phospholipids and phosphoproteins. Phosphorus also participates in the genetic machinery as the sugar phosphate backbone of nucleic acids and in nucleoproteins of chromatin. The abundance of phosphorus at multiple levels of cellular organization makes rigid control of plasma concentrations less important, and plasma fluctuations greater than 50 percent are physiologically tolerable.

Calcium and phosphate react to form insoluble tricalcium phosphate according to the equation

$$3Ca^{2+} + 2PO_4^{3-} \rightleftarrows Ca_3(PO_4)_2$$

From the law of mass action, we know that the concentration of calcium ions times the concentration of phosphorus divided by the concentration of tricalcium phosphate equals the **equilibrium constant, K_{eq}**:

$$[Ca^{2+}]^3[PO_4^{3-}]^2/[Ca_3(PO_4)_2] = K_{eq}$$

Since the concentration of the precipitate $Ca_3(PO_4)_2$ is also constant, a new constant called the **solubility product** can be defined as

$$K_{sp} = [Ca^{2+}]^3[PO_4^{3-}]^2$$

This equation says that insoluble calcium phosphate precipitates if the product of calcium ion concentration cubed and the phosphate ion concentration squared exceeds the K_{sp}. This means that calcium deposits can result from increased calcium concentration, increased phosphate concentration, or both. At physiological conditions, the solubility product equals 25. Since biological microenvironments are commonly altered by pH, ionic strength, or the presence of inhibitors, it is not uncommon for the calcium and phosphate product to range from 23 to 46 without precipitation. Since calcium is normally rigidly controlled, this variation is mostly related to phosphate fluctuations.

PHYSIOLOGY OF CALCIUM AND PHOSPHORUS

The average adult has 20 to 25 grams of calcium per kilogram body weight, or roughly 1,000 to 1,500 grams total. As mentioned previously, most of the calcium (99%) is in the skeleton, and approximately 1 percent circulates in the plasma and interstitial fluid. These two calcium pools are not isolated but are in rapid equilibrium exchange with the skeleton, which acts as a buffering depot not only for calcium ions but also for phosphate, magnesium, sodium, and most other ions. Approximately 4 to 6 grams of the total skeletal calcium is "exchangeable" or in rapid equilibrium with the plasma. This exchangeable calcium is 20 times greater than the total amount of plasma calcium.

Calcium balance is the sum of skeletal, gastrointestinal, and renal contributions (Fig. 8.1). The exact amounts vary with each individual, and approximate values are given below, but in a healthy person, calcium gains equal calcium losses. On the average, 500 milligrams of calcium enters the plasma daily from the skeleton through normal bone resorption, and another 500 milligrams goes back into the skeleton as new bone. Absorption by the stomach, duodenum, and ileum equals 300 milligrams, mainly from dairy products, meats, and greens (about 30 percent of the 1,000-milligram dietary intake is absorbed). Gastrointestinal losses are 150 milligrams daily from secretions and mucosal turnover. The kidneys filter 10,000 milligrams of calcium daily, but most is reabsorbed; daily urinary losses are 150 milligrams.

Despite the constant additions and subtractions, plasma calcium concentration remains stable, in the range of 8.5 to 10.1 milligrams per deciliter, which works out to a concentration of 4.5 to 5.6 millimolar. Children and adolescents may have values as high as 11 milligrams per deciliter. Calcium circulates in three forms: (1) protein bound, mainly to albumin, (2) organically complexed, mainly as calcium citrate, and (3) freely ionic as Ca^{2+}. About 45 percent of the circulating calcium is ionized Ca^{2+}, and this is the physiologically active form responsible for neuromuscular activity, blood clotting, etc. Forty percent is protein bound, and 15 percent is organically complexed; both are unavailable for biological reactions. A change in serum

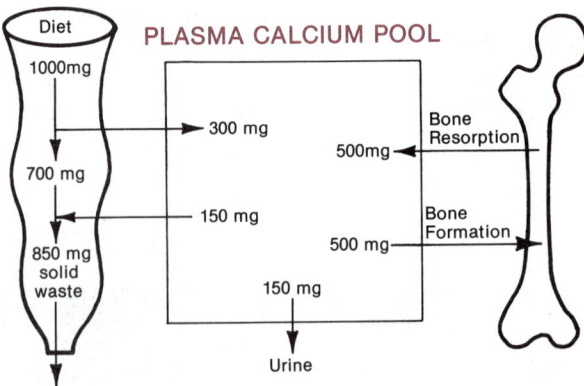

FIG. 8.1. Daily calcium balance in man. During an average day, 500 milligrams of calcium enters the plasma from bone resorption while an equal 500 milligrams reenter the skeleton in new bone formation. About 30 percent of the dietary calcium (300 milligrams) is absorbed, and 150 milligrams is lost from the GI tract. The kidneys filter 10,000 milligrams of calcium daily and reabsorb all but 150 milligrams which is excreted in the urine.

pH will shift the values of bound and ionic calcium. Acidosis increases the serum Ca^{2+} by decreasing protein binding (protons compete with Ca^{2+} for the binding sites); alkalosis decreases Ca^{2+} by increasing the binding to protein.

The skeleton also contains most of the inorganic phosphorus, normally about 80 to 85 percent. Muscle and other soft tissues contain 15 percent, and 1 percent circulates in the plasma, where levels are much more variable than calcium. Normal phosphorus values are 3.5 to 4.6 milligrams per deciliter, but variations of 3 to 4 milligrams per deciliter are common, particularly after a meal. Roughly 80 percent of plasma phosphate is dibasic HPO_4^{2-}, and 20 percent is monobasic $H_2PO_4^-$. Together, these two phosphates comprise the most important buffering system of the plasma.

Daily absorption of phosphorus from the gastrointestinal tract is about 2,000 milligrams, most of which is excreted in the urine. Combined renal filtration and secretion protects against hyperphosphatemia, and, as is discussed below, parathyroid hormone contributes to the protection by stimulating renal phosphate secretion.

HORMONAL REGULATION

Hormones guarantee a stable plasma calcium concentration in the face of constant bone turnover, dietary variations, and obligatory urinary losses. The three most important hormones for calcium and phosphorus metabolism are vitamin D, parathyroid hormone, and calcitonin.

Vitamin D

The word vitamin was coined by Funk at the turn of the century for certain "vital amines" required for health, above and beyond the full complement of protein, carbohydrate, fat, and minerals. Vitamin D is a lipid-soluble sterol hormone required for maintenance of calcium homeostasis; hypocalcemia occurs in its absence. Precursors of vitamin D come from two sources: (1) the diet and (2) the skin. Dietary vitamin D comes mainly from dairy products, liver, and fish; absorption occurs in the upper part of the small bowel. The integument is an organ of synthesis where sunlight (ultraviolet rays) photoconverts 7-dehydrocholesterol normally present in skin into vitamin D_3, also called cholecalciferol.

Cholecalciferol undergoes two hydroxylations, first in the liver and then in the kidney, to form the active metabolite, **1,25-dihydroxyvitamin D.** Liver hepatocytes contain an enzyme, **25-hydroxylase,** that converts cholecalciferol to 25-hydroxyvitamin D. Kidney cells in the proximal tubule contain a mitochondrial enzyme, **1-hydroxylase,** that converts 25-hydroxyvitamin D into **1,25-dihydroxycholecalciferol,** also called **calcitriol,** the

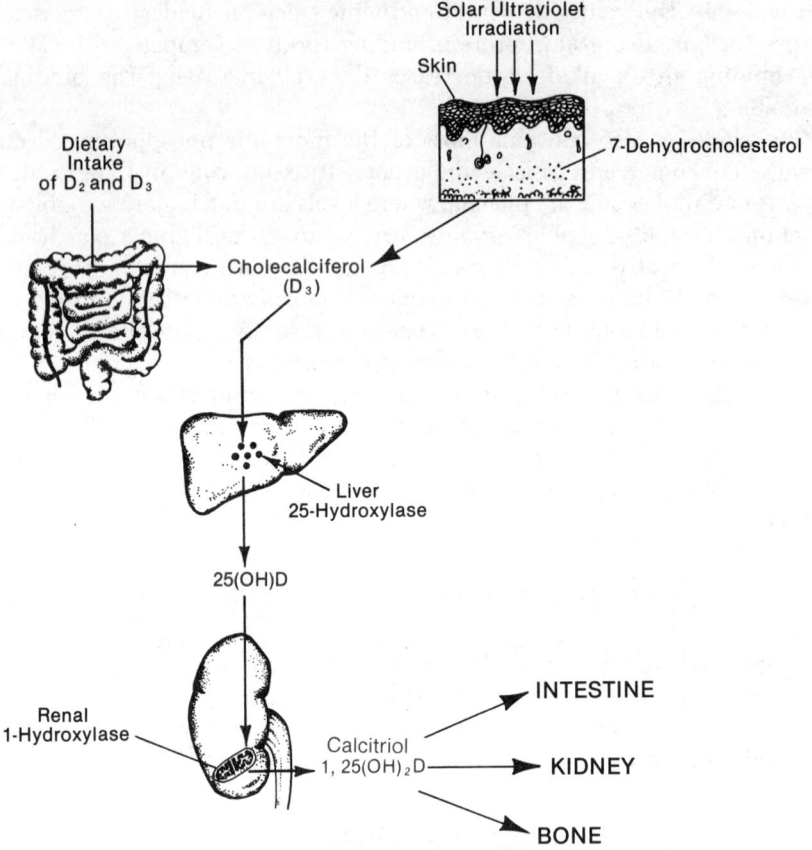

FIG. 8.2. Vitamin D metabolism. Cholecalciferol (D_3) comes from solar irradiation of cutaneous 7-dehydrocholesterol and from dietary intake. Liver 25-hydroxylation and renal 1-hydroxylation produce calcitriol (1,25-dihydroxycholecalciferol), the major active metabolite.

physiologically active form of vitamin D (Fig. 8.2). The normal range of calcitriol in the blood is 20 to 76 nanograms per milliliter. The control point for vitamin D metabolism is the renal enzyme 1-hydroxylase, an allosteric enzyme stimulated by parathyroid hormone and hypophosphatemia and inhibited by acidosis, calcitonin, and hyperphosphatemia.

The major organs affected by 1,25-dihydroxyvitamin D are the intestine and the skeleton. The effect on the intestine is to increase active transport of calcium by increasing the synthesis of messenger RNA for the **calcium binding protein.** Found in high concentrations at the brush border of the enterocyte, calcium binding protein moves calcium out of the gut, across the cell, and into the plasma.

1,25-dihydroxyvitamin D has two opposing effects on bone. The first and best-known effect is to facilitate mineralization of osteoid, preventing rickets and osteomalacia by supplying adequate amounts of calcium to mineralizing osteoid. The second effect is to work as an obligatory cofactor with parathyroid hormone in the efficient mobilization of calcium from bone.

Pathophysiology of vitamin D metabolism can arise from five possible situations: (1) decreased availability, (2) accelerated metabolism, (3) failure of the liver 25-hydroxylase, (4) failure of the kidney 1-hydroxylase, and (5) end-organ resistance (Table 8.5). Decreased availability can result from nutritional deficiency associated with decreased exposure to sunlight, as occurs in certain Asian populations or among the elderly and the infirm. Malabsorption syndromes disrupt the normal absorption of all lipid-soluble vitamins, contributing to decreased blood levels. Accelerated metabolism of vitamin D occurs in the liver when cytochrome P_{450}, a drug-metabolizing complex, is elevated. Certain drugs such as phenytoin, phenobarbital, and rifampin increase P_{450} levels and cause vitamin D metabolites to be degraded at an accelerated rate. Failure of the liver 25-hydroxylase occurs in certain liver diseases such as cirrhosis or after administration of the drug isoniazid, which competes with cholecalciferol for the 25-hydroxylase.

Failure of the 1-hydroxylase occurs in renal disease, renal tubular disorders, acidosis, and normal aging. In renal failure, when the creatinine clearance is between 25 and 30 milliliters per minute, enough nephrons have been lost to decrease the amount of 1-hydroxylase present and cause phosphate retention, which further inhibits the remaining enzyme. Renal tubular disorders include X-linked vitamin-D-resistant rickets and Fanconi syndrome. In X-linked vitamin-D-resistant rickets, patients have normocalcemia, hypophosphatemia, normal levels of vitamin D metabolites, and phosphaturia. In

TABLE 8.5. *Pathophysiologic mechanisms of vitamin D deficiency or resistance causing rickets or osteomalacia*

1. Decreased availability
    Low exposure to sunlight
    Low intake of vitamin D
    Malabsorption syndromes
2. Accelerated metabolism
    Use of phenytoin, phenobarbital, or rifampin
3. Failure of liver 25-hydroxylase
    Cirrhosis (alcoholic and primary biliary)
    Use of isoniazid
4. Failure of kidney 1-hydroxylase
    Renal failure
    Renal tubular disorders (X-linked resistant, Fanconi)
    Type I vitamin-D-resistant rickets
    Aging
5. End organ resistance
    Type II vitamin-D-resistant rickets

Fanconi syndrome, proximal tubular dysfunction contributes to 1-hydroxylase deficiency, and the kidneys lose many substances including amino acids, glucose, and phosphate.

Patients with **type I vitamin-D-resistant rickets** lack normal 1-hydroxylase levels. They have clinical rickets, hypocalcemia, hypophosphatemia, secondary hyperparathyroidism, normal levels of 25-hydroxyvitamin D, and low levels of 1,25-dihydroxyvitamin D. Treatment is either with massive doses of cholecalciferol or with the 1,25-dihydroxyvitamin D (calcitriol).

Patients with **type II vitamin-D-resistant rickets** have end organ resistance; that is, the target organs fail to respond to 1,25-dihydroxyvitamin D. Clinically, the patients have rickets, hypocalcemia, hypophosphatemia, secondary hyperparathyroidism, but normal vitamin D metabolite levels. Treatment is with large doses of 1,25-dihydroxyvitamin D, but because of end organ insensitivity, treatment is not fully effective. Table 8.5 summarizes the interrelationships of these diseases, and Table 8.6 shows the serum profiles in vitamin D abnormalities.

Parathyroid Hormone

The parathyroid glands are four small glands (6 × 3 millimeters) weighing about 30 milligrams located behind the thyroid gland. Derived from the third and fourth branchial pouches, the parathyroids contain two cell types, the chief cells, which produce parathyroid hormone (PTH), and the oxyphil cells, which appear to be senescent chief cells. Parathyroid hormone is released by the parathyroid glands in response to as little as 0.1 milligram per deciliter drop in plasma calcium. Parathyroid hormone controls the minute-to-minute levels of plasma calcium and protects against hypocalcemia.

Parathyroid hormone is a linear polypeptide hormone, 84 amino acids long, with a molecular weight of 9,500. Its synthesis is identical to that of all peptide hormones in that a large precursor molecule is made first. The initial peptide, assembled on the rough endoplasmic reticulum, has 115 amino acids and is called **"prepro-PTH."** A signal sequence of amino acids, the "pre" portion, shunts the nascent peptide into the endoplasmic reticulum cisternal space. Once inside the cisternal space and segregated from the cytosol, the "pre" segment is removed, converting the precursor to **"pro-PTH."** The "pro" segment is the molecular ticket for packaging into secretory granules in the Golgi apparatus. The principal secreted form of the hormone contains 84 amino acids, but the first 34 amino acids perform all known biological effects; the function of the remaining 50 amino acids is obscure (Table 8.7).

At the target organs, **cyclic AMP** mediates the effects of PTH. The major effect on the kidney is to increase distal tubular reabsorption of calcium and therefore decrease urinary calcium excretion. Parathyroid hormone also inhibits reabsorption of phosphate, causing increased urinary phosphate

TABLE 8.6. Serum profiles in vitamin D abnormalities

Condition	Serum calcium	Serum phosphate	Parathyroid hormone	25(OH)D	1,25(OH)$_2$D
Nutritional deficiency, malabsorption	Decreased	Decreased	Increased	Decreased	Decreased, normal, or sometimes elevated
Renal failure (creatine clearance < 30–50 ml/min)	Decreased	Increased	Increased	Normal	Decreased
Hypoparathyroidism	Decreased	Increased	Decreased	Normal	Decreased
X-linked hypophosphatemic rickets	Normal	Decreased	Normal	Normal	Decreased or normal
Vitamin-D-dependent rickets Type I (autosomal recessive)	Decreased or normal	Decreased	Increased	Normal	Decreased
Type II	Decreased	Decreased	Increased	Normal	Increased
Hyperparathyroidism	Increased	Decreased	Generally increased (overlap with normal)	Normal	Increased (considerable overlap with normal)
Hypervitaminosis D	Increased	Normal or increased	Decreased	Increased	Normal or increased

Adapted with permission from Audran and Kumar (1985).

TABLE 8.7. *Parathyroid hormone*

Molecule	Amino acid length	Function
Prepro-PTH	115	Segregation into the ER cisternal space
Pro-PTH	90	Ticket for Golgi processing and packaging
PTH	84	Principal secreted form
PTH_{1-34}	34	Full biological activity
PTH_{35-84}	50	Unknown

excretion (phosphaturia). Finally, PTH increases the activity of the 1-hydroxylase, increasing production of 1,25-dihydroxycholecalciferol.

Parathyroid hormone has both immediate and delayed effects on bone. The immediate effect is to increase extraction of calcium from the exchangeable skeletal pool. Both surface and deep osteocytes can mobilize calcium and phosphorus from the perilacunar bone surface and pump it into the plasma. Belanger named this process **"osteocytic osteolysis"**; in this way, bone makes an almost instantaneous contribution to plasma calcium. The delayed PTH effect is to increase osteoclast number, to increase the size of resorptive microvilli in the clear zone, and to increase the enzymatic activities of degradative hydrolases present in osteoclasts.

Pathology of PTH includes both hyperparathyroidism and hypoparathyroidism. Primary **hyperparathyroidism** affects mostly adults 20 to 40 years old. Eighty-five percent of the cases are caused by a parathyroid adenoma, 15 percent are caused by glandular hyperplasia, and rarely a parathyroid carcinoma causes hyperparathyroidism. Secondary hyperparathyroidism occurs in chronic renal insufficiency because the calcium loss through diseased kidneys and the phosphate retention both cause parathyroid hyperplasia. Subperiosteal bone resorption is the hallmark of hyperparathyroidism, along with osteopenia and intracortical osteoclastic tunneling. Patients have recurrent nephrolithiasis, peptic ulceration, mental changes, weakness, weight loss, nausea, vomiting, and polyuria. Malignancies producing PTH ectopically (pseudohyperparathyroidism) also cause hypercalcemia.

Hypoparathyroidism can be congenital or acquired. The most common acquired cause is postoperative hypoparathyroidism as a result of inadvertent removal of the glands or damage to their blood supply during thyroid surgery. Neuromuscular symptoms and signs predominate, including irritability, tetany, and convulsions. Clinical signs may include Chvostek's sign (tap on the nerve of facial muscles and get a grimace) and Trousseau's sign (blood pressure cuff elevation gives carpopedal spasm). Along with hypocalcemia, hyperphosphatemia occurs, and this accounts for the soft tissue calcifications. Therapy consists of oral calcium and vitamin D supplements.

Pseudohypoparathyroidism is a rare genetic disorder characterized by hypocalcemia, hyperphosphatemia, parathyroid hyperplasia, and elevated levels of PTH. The receptors for PTH at bone and kidney are abnormal, so the organs are blind to the hormone.

Calcitonin

The **parafollicular** or **C cells** of the thyroid gland produce calcitonin, a peptide hormone of 32 amino acids. C cells lie along the periphery and within the walls of thyroid follicles. Medullary thyroid carcinoma is a calcitonin-secreting tumor of C cells.

Hypercalcemia stimulates and hypocalcemia inhibits calcitonin secretion. It is an important hormone in fish that migrate from fresh to salt water and also in birds during the egg-laying cycle, but the exact physiological role in man is unknown; there are no known hormone excess or deficiency diseases associated with calcitonin. However, the pharmacologic effects of calcitonin on bone and kidney are known.

Calcitonin decreases the efflux of exchangeable calcium from bone and inhibits bone resorption by reducing osteoclast activity and numbers. At the kidneys, calcitonin increases both calcium and phosphorus excretion by blocking resorption. Calcitonin also stimulates renal formation of 24,25-dihydroxycholecalciferol, an inactive form of vitamin D, thus shunting precursor away from the 1,25-dihydroxyvitamin D pathway and decreasing 1,25-dihydroxyvitamin D concentration. These effects combine to produce a transient hypocalcemia and hypophosphatemia.

A currently attractive theory is that calcitonin acts in man to prevent postprandial hypercalcemia and to encourage a smooth incorporation of dietary calcium into the system. Another possibility is that calcitonin protects the skeletony from potential calcium depletion during pregnancy and lactation. Calcitonin has been used successfully in the treatment of some cases of hypercalcemia associated with Paget's disease.

METABOLIC BONE DISEASE

Osteopenia (Gk. *penia,* poverty, lack) is a generic term for the radiologic finding of a deficiency in bone radiodensity relative to normal age, sex, and race values. At least 30 percent of the bone mineral must be removed from a vertebral body or similar bone before osteopenia can be radiographically identified with certainty. Four histologic patterns of metabolic bone disease account for most cases of osteopenia: (1) osteoporosis, an absolute decrease in amount of bone; (2) osteomalacia and rickets, a failure of normal mineralization with a low ratio of mineral to matrix; (3) osteitis fibrosa, caused by hyperparathyroidism stimulating increased osteoclastic resorption; (4) marrow packing and malignancy, where bone mass is replaced by another tissue (Fig. 8.3) (Table 8.8).

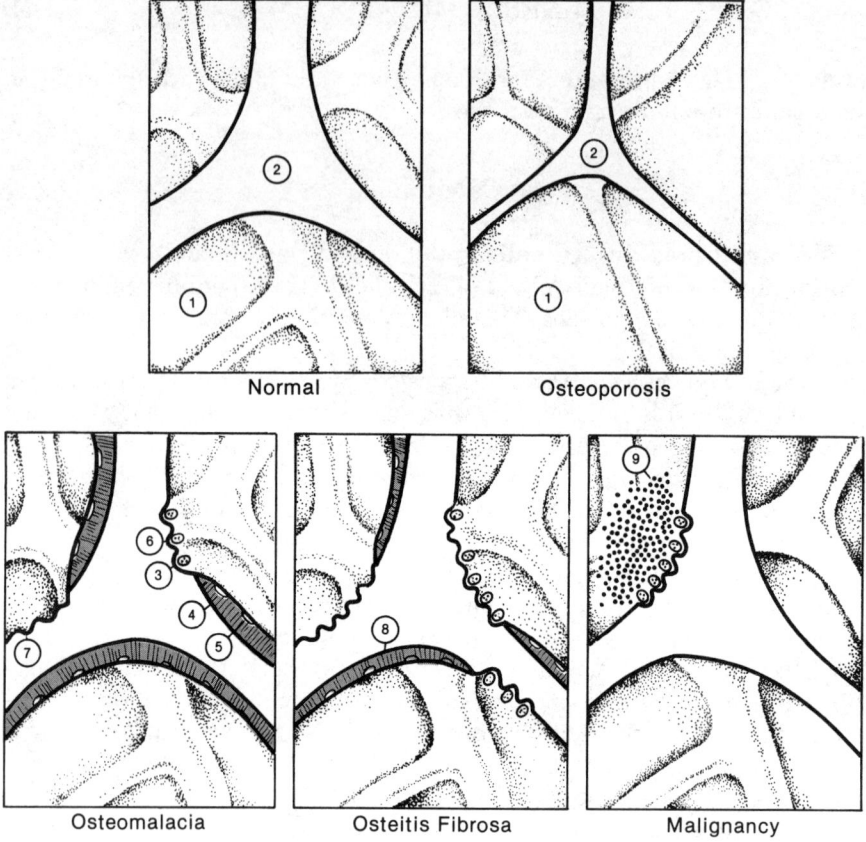

FIG. 8.3. Trabecular bone morphology in various osteopenias. In osteoporosis, mineral and matrix are decreased, but proportions are normal. In osteomalacia, mineralization is impaired, and the ratio of mineral to matrix is low, as also in osteitis fibrosa. In malignancy, bone resorption is increased through stimulation of osteoclasts, reducing both mineral and matrix. Key: 1, marrow cavity; 2, mineralized bone; 3, osteoclasts; 4, osteoid tissue; 5 osteoblast; 6, active bone-resorbing surface; 7, inactive bone-resorbing surface; 8, calcification front; 9, tumor cells. Reprinted with permission from Mundy (1978).

TABLE 8.8. *Serum chemistry values in different types of osteopenia*

Type of osteopenia	Serum calcium	Serum phosphate	Serum alkaline phosphatase	Serum PTH	Urine calcium
Osteoporosis	N	N	N	N	N or ↑
Osteomalacia	N or ↓	↓↓	↑	↑	↓
Osteitis fibrosa	↑	↓	↑	↑	↑
Renal osteodystrophy	N or ↓	↓	N or ↑	↑	N or ↓
Malignancy	N or ↑	↓	↑	N	N or ↑

Osteoporosis

Osteoporosis, the most common type of osteopenia, is a decrease in the amount of bone tissue in the skeleton, but the tissue that is present has a normal proportion of mineral to matrix; i.e., there is less bone per unit volume, but what is there is normal. Since the compressive strength of bone is proportional to the square of its apparent density, decreasing the density by a factor of 2 decreases the compressive strength by a factor of 4. Not suprisingly, the most common clinical problems associated with osteoporosis are bone pain and fracture. It is estimated that osteoporosis is responsible for 1.2 million fractures each year in the United States. The most common fractures in the year 1982 were 538,000 vertebral body compression fractures, 227,000 hip fractures, 172,000 distal forearm fractures, and 283,000 fractures at other limb sites.

Bone mass depends on skeletal growth and on the remodeling process, normally increasing until about the age of 30, when most people have reached their bone mass peak. Thereafter, bone mass remains constant until about 50, when it begins to decrease in both men and women, but the rate of decrease, and hence the magnitude of bone loss, is much greater in women. Men lose about 0.5 to 0.75 percent of their bone mass per year, but women lose 1.5 to 2 percent per year, and following menopause, the rate can approach 2.5 to 3 percent per year. The pathogenesis of osteoporosis involves an imbalance between bone formation and bone resorption, with resorption predominating. Table 8.9 lists some of the causes of osteoporosis, divided into primary and secondary osteoporosis.

Primary osteoporosis involves two syndromes of involutional osteoporosis: **type I (postmenopausal)** and **type II** (age related or **senile**) osteoporosis (Table 8.10). Type I osteoporosis is associated with estrogen deficiency in postmenopausal women and occasionally with androgen deficiency in men. The rate of trabecular bone loss is two to three times normal, and the rate of cortical loss is slightly above normal. These people sustain mostly vertebral compression fractures and distal forearm fractures.

Type II (senile) osteoporosis affects the entire aging population and involves both cortical and trabecular bone. The apparent reason for the resorption/formation imbalance is decreased osteoblast function and impaired production of 1,25-dihydroxycholecalciferol.

Certain risk factors for osteoporosis have been identified besides estrogen deficiency and aging, although premature menopause is still the most significant risk factor for women. Osteoporosis risk increases in women with a light skeletal structure such as small-boned Caucasian women of northern European extraction (of particular risk are women with slender build, fair complexion, freckles, and light hair). Osteoporosis is rare among black women, who tend to have more bone mass than Caucasian women.

TABLE 8.9 *Etiologies of osteoporosis*

Primary osteoporosis
  Type I (postmenopausal)
  Type II (age-related, "senile")
Secondary osteoporosis
  Endocrine disorders
    Hypercortisolism
    Hypogonadism
    Hyperthyroidism
  Decreased physical activity
    Immobilization
    Bed rest
    Weightlessness
  Drug induced
    Heparin therapy
    Steroid therapy
    Alcoholism
    Immunosuppressive therapy
  Idiopathic
    Transient osteoporosis
    Sudeck atrophy

Diet is also a risk factor in that low calcium intake throughout life predisposes to osteoporosis. The minimum daily calcium requirement is unknown, but the recommended daily requirement is 800 milligrams. The diet of many American women is unfavorable because it is low in calcium and high in

TABLE 8.10 *The two types of involutional osteoporosis*

	Type I (postmenopausal)	Type II (senile)
Age (years)	51–75	>70
Sex ratio (F:M)	6:1	2:1
Type of bone loss	Mainly trabecular	Trabecular and cortical
Rate of bone loss	Accelerated	Not accelerated
Fracture sites	Vertebrae (crush) and distal radius	Vertebrae (multiple wedge) and hip
Parathyroid function	Decreased	Increased
Calcium absorption	Decreased	Decreased
Metabolsim of 25(OH)D to 1,25(OH)$_2$D	Secondary decrease	Primary decrease
Main causes	Factors related to menopause	Factors related to aging

Adapted with permission from Riggs and Melton (1986).

phosphorus. For instance, a light breakfast, a leafy-green lunch, and a moderate dinner of meat, potatoes, vegetables, and salad rarely provides over 600 milligrams of calcium. The situation is aggravated if a soft drink substitutes for milk (skim milk = 250 milligrams/glass). Not only does the soft drink lack calcium, but the high phosphorus content increases osteoporosis (the postulated mechanism involves stimulation of PTH secretion). A calcium-poor diet, perhaps coupled with obligatory calcium losses of pregnancy and lactation, greatly increases the risk of osteoporosis.

The ideal therapy for primary osteoporosis is prevention, since established osteoporosis is so difficult to treat. A proper diet, regular exercise, and calcium supplements if necessary offer the best prophylaxis. Preventive recommendations for premenopausal women include at least 800 milligrams of calcium per day (either dietary or supplemental) and 1,500 milligrams per day after menopause because of decreased calcium absorption. Treatment of established osteoporosis includes a daily regimen of 1,500 milligrams of calcium, 400 units of vitamin D, plus sodium fluoride, 1 milligram per kilogram of body weight. In high-risk patients, addition of estrogen therapy can significantly reduce the incidence of fractures (either 0.625 milligrams conjugated estrogen or low-dose oral contraceptive).

Secondary osteoporosis is most commonly related to glucocorticoid excess, usually iatrogenic but occasionally resulting from Cushing disease. Steroids suppress intestinal calcium absorption and depress osteoblastic activity, thus favoring bone resorption over formation. Gonadal hormone deficiency leads to severe bone loss, but the exact mechanisms are obscure; apparently the normal levels of estrogens and testosterone somehow reduce the rate of bone resorption. Excess thyroid hormone stimulates osteoclastic bone resorption, and the loss is most rapid in skeletally immature patients, who have a high basal bone turnover rate.

Decreased physical activity, particularly immobilization, produces a rapid osteoporosis. The rate of bone resorption exceeds that of new bone formation because of decreased osteoblastic activity, and, as in most cases of secondary osteoporosis, bone loss is greatest in younger persons with high bone turnover rates.

Osteomalacia and Rickets

Normal mineralization requires a critical concentration of calcium and phosphorus ions, and below this concentration, mineralization is abnormal, resulting in osteomalacia and rickets. **Osteomalacia** occurs after skeletal maturity and is recognized by the accumulation of osteoid at sites of bone turnover. **Rickets** occurs in children and adolescents; not only does excessive osteoid accumulate, but so does unmineralized growth cartilage.

Children with rickets have skeletal deformities, muscular weakness, and hypotonia and are generally apathetic and irritable. Infants may have characteristic skull changes consisting of parietal flattening, prominence of the frontal bones (frontal bossing), and softening of many skull bones (craniotabes). Chest wall deformities include pectus carinatum, prominent costochondral junctions (rachitic rosary), and indentation of the lower rib cage at the insertion of the diaphragm (Harrison's groove). Abdominal muscles are weak; a potbelly and umbilical hernia are common. Legs as well as arms develop angular deformities, varus more commonly than valgus. Growth plate thickness increases as hypertrophic unmineralized cartilage accumulates, and the columns of cartilage cells become disordered. Joint reaction forces cause epiphyseal cupping, and failure of remodeling causes metaphyseal flaring. As a result, joints enlarge, and this is particularly obvious at the wrists, knees, and ankles. Radiographs show hazy, indistinct trabeculae, thin cortices, epiphyseal cupping, and flaring.

Osteomalacia most frequently presents with two findings, bone pain (often exquisitely painful to the touch) and diffuse proximal muscle weakness, commonly giving rise to a waddling gait. The radiographic picture is one of osteopenia, and in about 10 percent of the cases, radiographs reveal painless radiolucencies called **pseudofractures (Looser's zones).** Pseudofractures appear most frequently in the ribs, femoral neck, femoral shaft, pubic rami, and scapula. They are often bilaterally symmetrical, and many correspond to areas where nutrient arteries and veins go through the cortices. Two characteristic biochemical findings are elevated alkaline phosphatase and decreased plasma phosphorus. Calcium values can be normal or low, depending on the length of the disease process (normal early, low late). The histology of a bone biopsy shows an accumulation of increased amounts of unmineralized bone matrix (osteoid). Most patients who have osteomalacia will also have osteoporosis.

Common causes of osteomalacia are (1) vitamin D abnormalities, discussed above and listed in Table 8.6, (2) chronic hypophosphatemia, and (3) chronic metabolic acidosis. Conditions that result in chronic hypophosphatemia with normal vitamin D are familial hypophosphatemic rickets, some of the renal tubular disorders, and chronic phosphate-binding antacid use. Treatment involves only three therapeutic modalities: (1) vitamin D and calcium in most cases, (2) phosphate supplements in hypophosphatemia, and, finally, (3) sodium bicarbonate in acidosis.

Osteitis Fibrosa

Osteitis fibrosa cystica is a metabolic bone disease first recognized by von Recklinghausen as occurring in some patients with longstanding hyperparathyroidism. We have already discussed the pathophysiology of hyperparathyroidism. This section considers the skeletal and soft tissue lesions occurring in advanced hyperparathyroidism.

The skeleton in osteitis fibrosa has a generalized osteopenia, a lightly cellular fibrous replacement of the marrow spaces (osteitis fibrosa), and, occasionally, cystic lesions in cortical bone. The osteopenia resembles that of osteoporosis and osteomalacia, but focal osteoclastic resorption causes characteristic subperiosteal erosions best seen along the radial border of the middle phalanges, particularly along the middle index phalanx. The cortices are thin and have ragged, indistinct inner and outer surfaces. Osteoclasts may erode the distal clavicle and the lamina dura of the mandible. The tufts of the distal phalanges of the fingers and toes may also be absorbed. The skull has focal areas of decalcification surrounded by normal bone, giving a mottled "salt-and-pepper" appearance.

Brown tumors occasionally appear in patients with primary and secondary hyperparathyroidism. A brown tumor is a destructive, expanding, localized collection of fibrous tissue and giant cells. The tumor contains areas of cellular necrosis and liquefaction; hemorrhage accounts for the characteristic "brown" color. The histology of a brown tumor may at times be confused with a giant cell tumor, but the clinical presentation and the increased hemorrhage of a brown tumor help in the differentiation.

Soft tissue calcification is common, particularly in the kidney, where over 45 percent of patients have renal stones. Calcium also deposits in the cornea, producing band keratopathy. Approximately 5 percent of patients with hyperparathyroidism have peptic ulcer disease (because of increased gastrin activation), and 3 percent have pancreatitis.

Renal Osteodystrophy

Skeletal disease occurring in patients with chronic renal failure is a mixture of four histologic patterns: osteomalacia, osteitis fibrosa cystica, osteosclerosis, and osteoporosis. There is also soft tissue and vascular calcification.

The pathophysiology involves a disorder of mineral metabolism with uremia, defective vitamin D metabolism, and secondary hyperparathyroidism. Renal parenchymal disease results in the uremia and in low levels of the enzyme 1-hydroxylase. The deficiency of 1,25-dihydroxyvitamin D and the chronic uremia contribute to defective intestinal calcium absorption, generating hypocalcemia. Phosphate retention along with hypocalcemia causes secondary hyperparathyroidism.

Osteomalacia in chronic renal failure, particularly in patients with hemodialysis treatment, is caused not only by calcium abnormalities but also by aluminum toxicity. The aluminum comes from the dialysis fluid and also from oral consumption of aluminum-containing antacids given to control phosphate absorption.

Osteosclerosis occurs late in the course of renal failure and affects local areas, particularly the lower vertebrae. Dense bands of bone alternate with osteopenic bone, producing a **"rugger jersey spine."**

Laboratory findings in renal osteodystrophy include elevated BUN and creatinine. Plasma calcium is normal early but low in chronic disease. Phosphate, magnesium, alkaline phosphatase, acid phosphatase, and PTH levels are high. Urinary calcium is low. Serum protein electrophoresis is normal, but urine protein electrophoresis demonstrates nonspecific proteinuria secondary to the glomerular damage.

Marrow Packing Disorders and Malignancy

Marrow packing disorders and malignancy cause osteopenia by replacing normal bone tissue with other cells. For instance, hematopoietic precursors replace bone tissue in hyperplastic marrow disorders such as thalassemia and sickle cell disease. Gaucher cells containing glucocerebrosides fill the marrow spaces in Gaucher disease (a deficiency of the enzyme glucocerebrosidase).

Malignant plasma cells fill the marrow spaces and cause cortical and cancellous bone destruction in multiple myeloma, the most common adult primary bone tumor. The myeloma cells do not dissolve bone, but they produce osteoclast-activating factor, stimulating osteoclastic bone resorption, producing both a diffuse and a "punched-out" focal osteopenia. Patients are usually anemic, have proteinuria (Bence–Jones protein), and an "M" spike in the globulin region of a serum electrophoresis. Approximately 1 percent of myelomas are nonsecretory, and in these cases, a bone marrow biopsy is diagnostic.

ABNORMAL CALCIFICATION

Minerals may accumulate in soft tissues either as amorphous calcific deposits (calcification) or as ectopic bone (ossification). Calcification includes three types: (1) idiopathic, (2) metastatic, and (3) dystrophic. **Idiopathic calcifications** accompany normal aging; for instance, calcification of the costal cartilages occurs in 75% of adults. Metastatic and dystrophic calcifications are abnormal. Ossification includes two types: (1) heterotopic bone and (2) tumor bone (Table 8.11).

TABLE 8.11. *Soft tissue calcification and ossification*

Amorphous calcification	Ossification
(1) Idiopathic	(1) Heterotopic
(2) Metastatic	(2) Tumor
(3) Dystrophic	

TABLE 8.12. *Common causes of metastatic calcification*

Caused by hypercalcemia
 Hyperparathyroidism
 Metastatic bone disease
 Multiple myeloma
 Milk alkali syndrome
 Hypervitaminosis D
 Sarcoidosis
Caused by hyperphosphatemia
 Chronic renal failure
 Hypoparathyroidism
 Tumoral calcinosis

Metastatic calcification may occur in any condition with sustained elevation of either the plasma calcium or phosphate concentration. Calcification occurs when the product of the ionic concentrations in the tissue exceeds the solubility product constant of tricalcium phosphate. Tissues characteristically involved in metastatic calcification include the blood vessels, corneas, conjunctiva, viscera, skin, and subcutaneous tissue. Depositions are usually bilateral and fairly symmetrical. Table 8.12 lists the most common causes of metastatic calcification.

Dystrophic calcification occurs in areas of damaged tissue in the presence of normal levels of plasma calcium and phosphorus. The site of injury is the only area to calcify, so dystrophic calcifications are usually asymmetric. Table 8.13 lists the most common causes of dystrophic calcification.

The serum alkaline phosphatase is normal in dystrophic calcification, may be normal or elevated in metastatic calcification depending on the cause, and is elevated with early ectopic ossification from any cause. Bone scans are positive in all types of soft tissue calcification and ossification, and thus they are of no use in the differential diagnosis of these conditions, but they are

TABLE 8.13. *Common causes of dystrophic calcification*

Caused by acquired lesions
 Atherosclerosis
 Blunt trauma
 Chronic inflammation
 Dermatomyositis
 Injection sites
 Scleroderma
 Thermal injury
 Tumor calcification
Caused by inherited disorders
 Alkaptonuria
 Ehlers–Danlos syndrome
 Homocystinuria
 Pseudoxanthoma elasticum
---

useful for assessing the extent of involvement and the maturity of the ectopic bone.

Skeletal muscle ossification after trauma (myositis ossificans) is the most common cause of heterotopic bone. Certain ethnic groups are prone to nontraumatic ectopic bone, such as calcification of the posterior longitudinal ligament in the Japanese. Ossification differs from calcification in that with ossification the histology discloses the presence of osteocytes, and the radiograph demonstrates trabeculation (see Chapter 4 and Table 4.5).

GLOSSARY

Calcitonin A 32-amino-acid polypeptide hormone released by the parafollicular cells of the thyroid in response to hypercalcemia. The exact physiological role is unknown, but it is thought to protect against postprandial hypercalcemia.

Calcitriol 1,25-Dihydroxycholecalciferol, also known as 1,25-dihydroxyvitamin D.

Calcium binding protein A protein found in high concentrations at the brush border of intestinal epithelial cells that moves calcium out of the gut, across the cell, and into the plasma. 1,25-Dihydroxyvitamin D is necessary for the synthesis of calcium binding protein.

Cholecalciferol The unhydroxylated precursor of vitamin D, also called vitamin D_3.

1,25-Dihydroxycholecalciferol Also called calcitriol and 1,25-dihydroxyvitamin D, the physiologically active form of vitamin D, which is synthesized from 25-hydroxyvitamin D in the kidney by the mitochondrial enzyme 1-hydroxylase.

Dystrophic calcification Deposition of amorphous calcium salts in areas of damaged tissue, occurring in the presence of normal levels of plasma calcium and phosphorus.

Ectopic ossification Also called heterotopic ossification, the formation of extraskeletal bone tissue such as posttraumatic myositis ossificans.

1-Hydroxylase The controlling enzyme of vitamin D metabolism, located in renal mitochondria. It is an allosteric enzyme stimulated by PTH and hypophosphatemia and inhibited by acidosis, calcitonin, and hyperphosphatemia.

Ionic radii The size, usually expressed in angstroms or picometers, of an atom without its valence ions; K^+ = 133 picometers, Ca^{2+} = 99 picometers, Na^+ = 95 picometers.

Looser's zones Also called pseudofractures, painless radiolucencies that look like unicortical fractures, occurring in osteomalacia.

Metastatic calcification Calcification occurring in normal tissues as a result of a sustained elevation of either the plasma calcium or the phosphorus concentration.

Osteitis fibrosa The metabolic bone disease caused by excess, prolonged hyperparathyroidism, characterized by fibrous replacement of the marrow spaces, cystic lesions of the cortex, and occasional brown tumors.

Osteocytic osteolysis A term coined by Belanger to describe the immediate mobilization of the exchangeable skeletal calcium by osteocytes. This process accounts for the almost instantaneous response of the skeleton to PTH.

Osteomalacia Metabolic bone disease occurring after skeletal maturity, characterized by the accumulation of unmineralized osteoid at the sites of bone turnover.

Osteopenia The radiological finding of a deficiency in bone mass relative to normal age, sex, and race values.

Osteoporosis A decrease in the amount of bone tissue in the skeleton, but the remaining tissue has a normal proportion of mineral to matrix.

Parathyroid hormone An 84-amino-acid polypeptide hormone made by the chief cells of the parathyroid glands that guards against hypocalcemia by three mechanisms: (1) decreasing the renal fractional excretion of Ca^{2+}, (2) mobilizing the exchangeable calcium from the skeleton, (3) stimulating renal 1-hydroxylase and thus increasing vitamin D concentration.

Prepro-PTH The first precursor of parathyroid hormone, 115 amino acids long, synthesized at the endoplasmic reticulum. The function of the "pre" portion is to shunt the nascent peptide out of the cytosol and into the ER cisternal space.

Primary hyperparathyroidism Increased PTH secretion because of an abnormality of the parathyroid glands such as an adenoma, glandular hyperplasia, or a parathyroid carcinoma.

Pro-PTH The direct precursor of PTH, 90 amino acids long. The function of the "pro" portion is to direct the packaging of PTH into secretory granules in the Golgi apparatus.

Pseudohyperparathyroidism That condition arising from the ectopic production of PTH by a malignancy.

Pseudohypoparathyroidism End organ resistance to PTH, thought to be caused by an abnormal PTH receptor protein. This is a genetic disorder characterized by hypocalcemia, hyperphosphatemia, parathyroid hyperplasia, and elevated plasma levels of PTH.

Renal osteodystrophy Skeletal disease occurring in patients with chronic renal failure (uremia, hyperphosphatemia, hypocalcemia) that is a mixture of osteomalacia, osteitis fibrosa, osteosclerosis, and osteoporosis.

Rickets Osteomalacia in children, characterized by accumulation of excess unmineralized osteoid and also unmineralized growth cartilage, giving rise to skeletal deformities.

Rugger jersey spine Roentgenographic picture of vertebral bodies in advanced renal osteodystrophy, showing alternating bands of osteosclerosis and osteopenia.

Secondary hyperparathyroidism Increased PTH secretion by normal parathyroid glands under the influence of chronic hypocalcemia or hyperphosphatemia, such as in chronic renal insufficiency.

Type I involutional osteoporosis Caused by factors related mainly to menopause, involves trabecular bone with vertebral and distal radial fractures. Calcium metabolism and parathyroid function are decreased.

Type II involutional osteoporosis Caused by factors related mainly to aging, involves cortical and trabecular bone. Calcium absorption is decreased, and parathyroid function is increased.

Type I vitamin-D-resistant rickets Abnormal vitamin D metabolism because of a decrease in renal 1-hydroxylase enzyme, inherited as autosomal recessive.

Type II vitamin-D-resistant rickets End organ resistance to normal vitamin D.

BIBLIOGRAPHY

Audran, M., and Kumar, R. (1985): The physiology and pathophysiology of vitamin D. *Mayo Clin. Proc.*, 60:851–866.
Austin, L. A., and Heath, H. (1981): Calcitonin. Physiology and pathophysiology. *N. Engl. J. Med.*, 304:169–278.
Connor, J. M. (1983): *Soft Tissue Ossification*, pp. 1–15. Springer-Verlag, New York.
Goldsmith, R. S. (1969): Hyperparathyroidism. *N. Engl. J. Med.*, 181:367–374.
Hahn, T. J. (1986): Physiology of bone: Mechanisms of osteopenic disorders. *Hosp. Pract.*, 21:73–90.
Kaplan, F. S. (1983): Osteoporosis. *Ciba Clin. Symp.* 35:1–32.
Klein, K. L., and Maxwell, M. H. (1984): Renal osteodystrophy. *Orthop. Clin. North Am.*, 15:687–695.
Lane, J. M., and Vigorita, V. J. (1983): Osteoporosis. *J. Bone Joint Surg.*, 65A:274–278.
Mundy, G. R. (1978): Differential diagnosis of osteopenia. *Hosp. Pract.*, 14:65–72.
Nordin, B. E. C. (1984): *Metabolic Bone and Stone Disease*, pp. 112–122. Churchill Livingstone, New York.
Riggs, L., and Melton, L. J. (1986): Involutional osteoporosis. *N. Eng. J. Med.*, 314:1676–1684.
Rosenblatt, M. (1982): Pre-proparathyroid hormone, proparathyroid hormone, and parathyroid hormone. The biological role of hormone structure. *Clin. Orthop.*, 170:160–275.
Spencer, H. (1982): Osteoporosis: Goals of therapy. *Hosp. Pract.*, 17:131–147.
Tannenbaum, H. (1984): Osteopenia in rheumatology practice: Pathogenesis and therapy. *Semin. Arthritis Rheum.*, 13:337–348.

9
Inflammation, Immunology, Healing

INFLAMMATION

Inflammation is a general tissue response to injury (L. *inflamare,* to burn), and the response is relatively independent of the nature of the damaging agent. Inflammation is one of the most important host defense mechanisms because it initiates the attack on the injurious agent, and it sets the stage for repair. The **cardinal signs** of inflammation are **redness, swelling, heat,** and **pain** (summarized by the Roman encyclopedist Celsus, 30 B.C. to 38 A.D., as *rubor et tumour cum calore et dolore*). In 1858, Virchow added a fifth sign of inflammation, loss of function. The morphological changes responsible for these clinical signs are local vasodilation, extravascular accumulation of plasma and leukocytes, cellular necrosis with release of chemical mediators.

Mediators of Inflammation

Tissue injury causes the release of **vasoactive substances** and **chemotactic factors** (chemical mediators) that produce hemodynamic changes, increase vascular permeability, and stimulate leukocytic migration. As a result of increased vasopermeability, plasma proteins, including globulins and fibrinogen, leak into the tissue where fibrinogen is converted into a fibrin meshwork. Chemotactic factors, including complement peptide fragments, C3a, and C5a, stimulate polymorphonuclear leukocytes to escape from capillaries and accumulate in the area (Table 9.1).

Vasoactive substances relax perivascular smooth muscle cells, increasing blood flow through the injured area by as much as tenfold (the basis of *rubor* and *calor*). Vasoactive substances also affect actin fibers within endothelial cells, leading to dilation of capillaries and opening of junctions between the endothelial cells through which plasma escapes into the tissue space (the

TABLE 9.1. *The inflammatory reaction*

Time	Mediator	Site	Hemodynamics	Permeability	WBC changes	Appearance
Immediate	Neurogenic	Arterioles	Vasoconstriction	None	None	Blanching
Early (5–30 minutes)	Histamine Serotonin	Arterioles	Vasodilitation	Increased	None	Redness Swelling Heat Pain
	Prostaglandins Complement Bradykinin	Capillaries	Open	Increased	None	As above
	Other mediators	Venules	Engorgement	Increased	Pavementing Adhesion	As above
Late (30–60 minutes)	Chemotactic factors	Capillaries	Engorgement	Increased	Emigration Diapedesis	As above

Adapted from Angell and Robbins (1979).

INFLAMMATION, IMMUNOLOGY, HEALING

basis of *tumour*). Histamine and serotonin are the major vasoactive amines; other vasoactive substances include the complement components C3a, C5a, and the oligopeptide bradykinin. **Bradykinin,** composed of nine amino acids, is the most potent pain-producing chemical known (the basis of *dolor*).

Perhaps the most versatile vasoactive substances are **prostaglandins (PGs).** Derived from fatty acids of cell membranes, PGs are five-membered ring structures. Substitutions on the ring are the basis for designation by the letters E and F. Prostaglandins of the 2 series are all derived from arachidonic acid and have two double bonds in their side chains, therefore the subscript 2, e.g., PGE_2. A closely related compound, thromboxane A_2, has a six-membered ring structure.

Cells do not store PGs but rapidly synthesize them in response to many different stimuli, tissue injury being one example. The enzyme **cyclooxygenase** converts plasma membrane arachidonic acid into an endoperoxide, PGH_2, and isomerases then convert PGH_2 into other PGs or into thromboxane. The tissue distribution of various isomerases determines the type of PGs that can be synthesized, and different tissues have quite different isomerases. For example, platelets contain the isomerase to make thromboxane A_2, whereas endothelial cells have the isomerase to make PGI_2 (prostacyclin), and synovial tissue synthesizes mostly PGE_2.

Prostaglandins have potent physiological effects, but the effects of different PGs differ widely. For instance, thromboxane A_2 causes platelet aggregation and vasoconstriction, whereas PGI_2 and PGE_2 are vasodilators and act synergistically with histamine and bradykinin. Prostacyclin and PGE_2 work via cell membrane receptors that are linked to the enzyme adenylate cyclase, increasing the intracellular concentration of cyclic AMP in target cells. Thromboxane A_2, the antagonist of PGE_2, increases the concentration of cyclic GMP, the intracellular antagonist of cyclic AMP.

Leukotrienes are related closely to PGs but are derived from arachidonic acid by the enzyme **lipoxygenase,** an enzyme found in high concentrations in neutrophils. Leukotrienes are powerful proinflammatory agents, acting as very potent chemotactic factors, increasing vascular permeability, and causing the release of PMN lysosomal enzymes into the site of inflammation.

Antiinflammatory Medications

Nonsteroidal antiinflammatory drugs, **NSAIDs** (aspirin, indomethacin, phenylalkanoic acid derivatives, etc.), act by inhibiting the biosynthesis of prostaglandins, blocking the action of the rate-limiting enzyme, cyclooxygenase. Aspirin irreversibly inhibits cyclooxygenase by acetylating the enzyme. Other NSAIDs reversibly inhibit the enzyme, so their action is dose dependent, requiring adequate and continuous plasma concentration of the drug.

Glucocorticoid steroids are powerful inhibitors of the inflammatory reaction (dexamethasone > prednisolone > hydrocortisone). The exact mechanism is unknown, but current research supports the theory that glucocorticoids inhibit the release of arachidonic acid from plasma membrane phospholipids. Other drugs such as gold compounds, penicillamine, and colchicine have no effect on PG synthesis but inhibit inflammation by other cell-specific effects (i.e., colchicine inhibits lymphocyte proliferation).

Chronic Inflammation

The predominant cells of acute inflammation are polymorphonuclear leukocytes and macrophages; the acute reaction is characterized by vascular and exudative changes. If inflammation continues for 4 to 6 weeks, it eventually becomes a chronic inflammation. The cellular reaction of chronic inflammation is pleomorphic, the major cell types being macrophages, lymphocytes, plasma cells, giant cells, and fibroblasts. Both cell proliferation and cell destruction occur in chronic inflammation. Proliferation involves chiefly fibroblasts and capillary buds. Cell destruction continuously generates lysosomal enzymes and other chemical mediators that perpetuate the inflammatory reaction and continue tissue damage.

Many events can lead to chronic inflammation. For instance, an acute inflammatory reaction may simply fail to subside and become chronic. Certain microorganisms such as those that cause tuberculosis, leprosy, syphilis, and actinomycosis characteristically induce a chronic inflammatory reaction. Also, repeated physical trauma or the presence of chemical irritants such as calcific deposits or calcium-containing crystals can prolong and perpetuate inflammation. Disturbances in the immune system may contribute to certain types of chronic inflammation such as rheumatoid arthritis, ulcerative colitis, and chronic hepatitis.

IMMUNOLOGY

The immune system allows the body to recognize, inactivate, and destroy foreign materials. It can also recognize "self" molecules and differentiate them from "nonself"; failure to make this distinction and reaction with host cells is the basis of autoimmune reactions. Immune system response can be considered to involve three phases (1) **recognition** of foreign material, (2) **proliferation** of immune cells (lymphocytes or plasma cells), and (3) **interaction** with the immunogen. The afferent arm of the immune system is responsible for recognition and determination if a material is foreign. The efferent arm carries out the immune interaction, resulting in inactivation or destruction of the foreign material.

Genes located on the sixth chromosome within the **HLA** region control the immune system. Products of these genes control antigen recognition, antibody production, lymphocyte proliferation, cytotoxicity, and immune suppression. These genes direct two types of immune response, **humoral** responses that involve the production of antibodies and **cell-mediated** immune responses that involve production of specialized cells that either **kill** other cells, **suppress** the immune response, or **help** out other cells.

Humoral Immunity

An **antibody (immunoglobulin)** is a globular protein synthesized by a plasma cell in response to the presence of a foreign material (antigen). Small foreign molecules (haptens) also elicit antibody production if they are attached to larger carrier molecules. Antibody has a specific affinity for the foreign material that elicited its synthesis, and the specificity of the antibody is directed against a particular molecular site on the antigen called the antigenic determinant.

Plasma cells are derived from **B lymphocytes** (named for the organ in birds, the bursa of Fabricius, necessary for maturation); in humans, B cells originate from stem cells in the bone marrow. B cells have specific immunoglobulins as an integral part of the cell surface.

An antigen will bind to and activate only one particular B cell, the one that contains surface antibody against the antigen. That particular B cell then proliferates and differentiates into many plasma cells, which synthesize identical immunoglobulins that neutralize the antigen.

All immunoglobulins consist of light chains (23,000 daltons) and heavy chains (50,000 daltons) with a variable region (about 108 amino acid residues at the amino-terminal end) and a constant region. Antigen-binding sites are formed by amino acids within the variable regions, and special functions such as complement activation occur at the constant region. There are five known types of immunoglobulins, which make up about 20 percent of the plasma proteins by weight: IgM, IgG, IgA, IgD, and IgE.

Immunoglobulin G accounts for 75 to 80 percent of all immunoglobulins and consists of two identical heavy chains and two identical light chains arranged as a Y-shaped protein (Fig. 9.1). Proteolytic enzymes split IgG into a **Fab** (fragment *a*ntigen *b*inding) region and a **Fc** (*c*rystallizes) region. The Fab fragment contains the hypervariable amino acid sequences that function in antigen binding. The Fc region binds to phagocytic cells, thereby increasing the efficiency of phagocytosis and also activating the first component of the complement system. The IgG immunoglobulins are the only antibodies to cross the placenta.

Immunoglobulin M is the largest antibody, a pentamer of five identical subunits plus a nonimmunoglobulin chain, the J chain (Fig. 9.1). It is the first antibody produced against any new immunogen, but it accounts for only

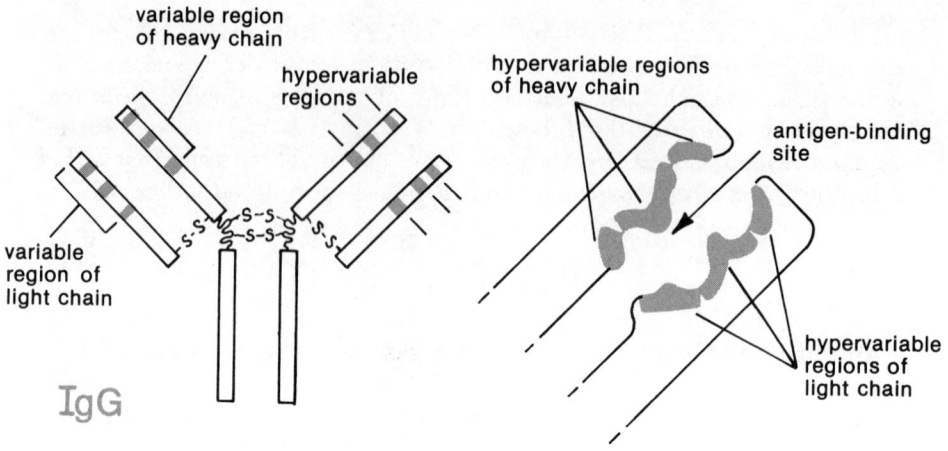

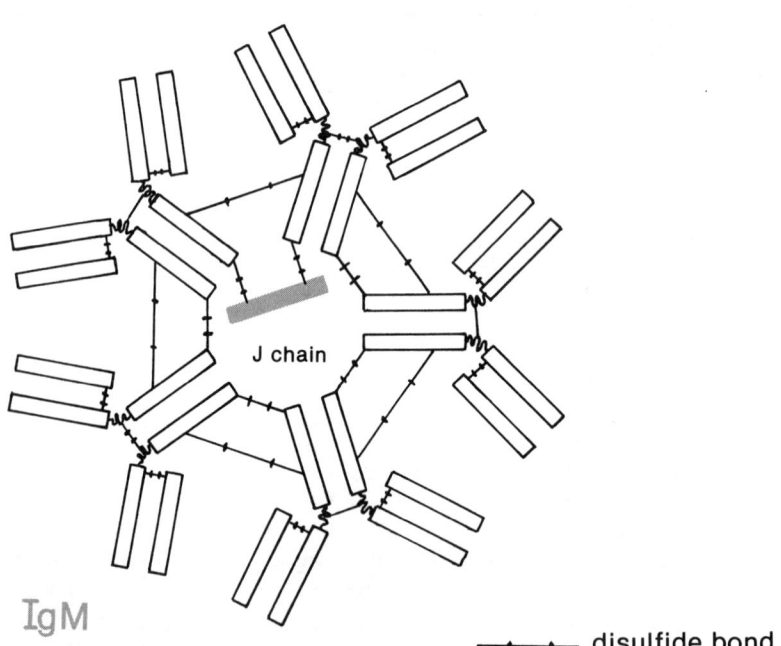

FIG. 9.1. Schematic drawing of the structures of IgG and IgM. Three hypervariable regions in each light and heavy chain make the two antigen-binding sites; IgM has five subunits plus a special J chain held together by disulfide bonds. Adapted with permission from Albers et al. (1983).

5 to 10 percent of the total blood immunoglobulin. As the immune response gets underway, immunoglobulin production shifts to IgG antibody. Because of its large size, IgM is particularly effective in agglutination reactions against bacteria and other macromolecular antigens. It is also more effective than IgG in activating the complement system.

Immunoglobulin A is the major antibody in biological secretions (intestinal, respiratory, saliva, tears, and milk), accounting for 10 to 15 percent of blood immunoglobulins. **Immunoglobulin E** molecules bind to histamine-containing basophils and mast cells, participating in allergic reactions such as asthma, hives, and hay fever. Finally, the exact role of **IgD** is unknown.

In routine immunization, the antibodies are to a certain extent heterogeneous in class and in specificity, so their use as molecular probes or as diagnostic agents is limited. **Monoclonal antibodies** differ from the usual antibodies in that they are completely homogeneous, consisting of absolutely identical immunoglobulins. Monoclonal antibodies are obtained in two ways: (1) from plasma cells of patients with multiple myeloma and (2) from hybridoma cells produced by fusing a lymphocyte that can make the desired antibody with a myeloma cell. The resulting hybrid grows in cell culture and synthesizes only one absolutely specific antibody. The absolute specificity and ease of preparation of monoclonal antibodies make them valuable diagnostic, therapeutic, and research agents.

Complement is another component of humoral immunity that "complements" the action of antibodies. Complement is actually a system of 18 plasma proteins that, after activation, interact sequentially to produce inflammatory effects, membrane lysis, and bacterial killing. Each complement component is designated by the letter "C" followed by a number of the component; lower-case letters following the numbers indicate activated components, such as C1q, or C5a. The activity and plasma concentration of complement are not affected by immunization, unlike those of the immunoglobulins.

After reacting with their antigens, IgG and IgM activate complement by what is known as the classical pathway. Other antigens such as bacterial cell envelopes activate complement by the alternate pathway. Active C3 (C3b) binds to bacteria (opsonization) to facilitate phagocytosis. C3b also initiates assembly of a lytic complement complex, made up of other complement components, that when assembled disrupts membrane integrity and kills cells. C5a is chemotactic for PMNs and monocytes, as are some of the other peptides split off during complement activation.

Cell-Mediated Immunity

Three different subpopulations of T cells carry out cell-mediated immunity: (1) cytotoxic T cells, (2) helper T cells, and (3) suppressor T cells. **Cyto-**

toxic T cells bind to and kill virus-infected host cells or any foreign cell. Helper and suppressor T cells act as regulators of the immune response. **Helper T cells** do not secrete antibody, but they are essential for B-cell antibody response to the presence of antigen. They also help cytotoxic T cells and activate macrophages by releasing lymphokines (interleukin 2 and MIF, migration inhibition factor). **Suppressor T cells** suppress the response of B cells or other T cells and are responsible for immunological tolerance. Approximately 70 percent of the blood lymphocytes are helper T cells, and the remaining 30 percent are cytotoxic and suppressor T cells.

Most lymphocytes continuously recirculate through the blood and lymph. They leave the bloodstream by squeezing out between endothelial cells, and after moving through the tissue spaces, they reenter small lymphatic vessels and collect downstream in lymph nodes. Eventually, lymphocytes enter the thoracic duct, which carries them back into the bloodstream.

T cells are responsible for graft rejection. Both cytotoxic and helper T cells attack and destroy an allograft or xenograft unless the immune system is suppressed with drugs. Immunodeficient patients who receive bone marrow allografts can experience rejection of the recipient tissue by the graft, the graft-versus-host response, which can be fatal. In transplantation reactions, the T cells direct their response against cells marked with foreign cell-surface antigens or histocompatibility antigens.

Major Histocompatibility Antigens

Histocompatibility antigens are cell-surface glycoproteins (antigens) encoded by a group of genes located on the short arm of chromosome 6 called the **major histocompatibility complex (MHC).** These genes ultimately control the interactions of T cells, B cells, and macrophages. The MHC antigens are also called **HLA antigens** (*h*uman *l*eukocyte-*a*ssociated antigens) because they were first found on leukocytes.

There are two classes of cell-surface MHC antigens, class I and class II. Most cells, including platelets, have class I antigens, which are coded by three loci within the MHC complex, the HLA-A, HLA-B, and HLA-C regions. **Class I antigens** are important because cytotoxic T cells respond to foreign class I antigens. **Class II antigens** have a much narrower tissue distribution, being expressed only in cells concerned with immune responses, i.e., most B cells, some T cells, and some macrophages. They are coded by the HLA-D locus, are essential for antigen presentation to immunocompetent cells, and seem to control the intensity of the immune response.

When a virus infects a cell, some of the viral antigen will migrate to the cell membrane and associate with class I MHC glycoproteins; this combination of host class I antigen plus viral antigen tells cytotoxic T cells to kill the cell (along with the infecting virus). There is such a wide distribution of

TABLE 9.2. *Diseases associated with certain HLA antigens*

Diseases	HLA association
Seronegative spondyloarthropathies	
Ankylosing spondylitis	B 27
Reiter syndrome	B 27
Psoriatic arthritis	B 27
Autoimmune diseases	
Rheumatoid arthritis	DR 4
Systemic lupus erythematosis	DR 3
Graves disease	DR 3
Sjogren syndrome	B 8

class I antigens because so many different cells can potentially be infected with viruses and need this defense mechanism.

Class II antigens work in a different way. It is thought that bacterial antigens directly associate with class II antigens, thereby signaling helper T cells, which in turn activate B cells and macrophages, leading to antibody production and binding, complement fixation, and ultimately phagocytosis.

Certain rheumatic diseases and autoimmune diseases (over 50 diseases so far) are associated with either **HLA-B27** or a variety of **HLA-DR** subtypes (Table 9.2). The HLA-DR class II glycoproteins are thought in most cases to predispose individuals to particular autoimmune diseases (again the exact mechanism is obscure). HLA-DR4 predisposes to rheumatoid arthritis, and HLA-DR3 to systemic lupus erythematosis. The majority of patients (>95%) with ankylosing spondylitis have HLA-B27 antigens (found in only 4% to 7% of the general population). Thus, assays for B-27 offer practical diagnostic assistance in the work-up of ankylosing spondylitis. In fact, all of the spondyloarthropathies (including Reiter syndrome, psoriasis, inflammatory bowel syndrome) have a strong association with HLA-B27. Current theories suggest that the pathogenesis of these disease associations involves abnormalities in the recognition phase of helper and cytotoxic T cells.

CRYSTAL-INDUCED INFLAMMATORY ARTHRITIS

This section reviews the three most prevalent crystal-induced inflammatory diseases: (1) gout, (2) CPPD crystal disease (pseudogout, chondrocalcinosis), and (3) basic calcium phosphate (BCP) crystal disease (Table 9.3). The acute clinical manifestations of these crystal-induced diseases depend on crystal phagocytosis by polymorphonuclear leukocytes. Once inside a leukocyte, the crystal causes cell disintegration, releasing the crystal and all the cellular contents (including lysosomal enzymes) into the joint space, accel-

TABLE 9.3. *Crystal deposition disease*

Disease	Crystal
Gout	Monosodium urate
CCPD diesase (pseudogout, chondrocalcinosis)	Calcium pyrophosphate dihydrate
BCP disease (calcific bursitis, etc.)	Hydroxyapatite
	Tricalcium phosphate
	Octacalcium phosphate

erating tissue damage, and perpetuating the inflammatory response (Fig. 9.2).

Gout

Monosodium urate (MSU) crystals precipitated in joints or soft tissues cause gout (L. *gutta,* a drop, reflecting the ancient belief of a malevolent humor dropping into weakened joints). Uric acid is the end product of purine metabolism in man, unlike other mammals, because he lacks uricase (an enzyme that permits further degradation of uric acid). Unfortunately, uric acid is relatively insoluble, and serum is saturated at about 7 milligrams per deciliter; higher values represent hyperuricemia, the biochemical hallmark of gout. Hyperuricemia results when there is either an excess production or a decreased excretion of uric acid.

Acute gouty arthritis tends to affect the lower extremities, particularly the first metatarsophalangeal joint, which becomes swollen and exquisitely painful **(podagra).** Attacks are so sudden and so severe that they may be mistaken for cellulitis or septic arthritis. Definitive diagnosis of gouty arthritis requires identification of MSU crystals in synovial fluid. The synovial fluid appears milky and opaque with a high white cell count (often >15,000), but the glucose is normal, and the Gram stain is negative. Crystals of MSU are needle-shaped and are best seen by means of polarized light (crystals are yellow when their long axis is parallel to that of the microscope's red compensator). A presumptive rather than a definitive diagnosis can be made if a patient has a classic history of acute monoarticular arthritis and hyperuricemia and responds to medication.

Untreated gout passes through three clinical phases: (1) acute gouty arthritis, (2) a quiescent period of interval gout, and, finally (3) chronic tophaceous gout (*tophi,* Gr. chalk stone). **Tophi** consist of a nodular core of MSU crystals surrounded by a fibrotic, chronic inflammatory reaction. They accumulate in periarticular and subcutaneous tissue, commonly along the Achilles tendon, in the olecranon bursa, or in the synovium. Occasionally

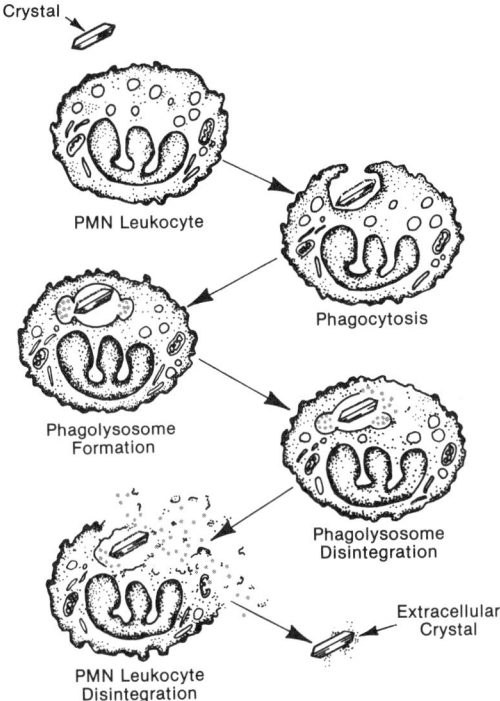

FIG. 9.2. The events of crystal-induced leukocyte destruction. Crystals phagocytosed by PMN leukocytes become engulfed in phagosomes, which fuse with lysosomes. The crystal destroys the phagolysosomal membrane, and enzymes leak into the cytosol, producing autolysis and cell death. Cell contents and the crystal spill into the joint to amplify the inflammation. Reprinted with permission from McCarty (1986).

deposits appear in the helix or the antihelix of the ear. Longstanding gout predisposes people to gouty nephropathy from precipitation of MSU crystals in the renal interstitium or in the collecting tubules. Avascular necrosis of the femoral head also occurs with increased frequency.

Treatment of gout can be divided into management of the acute flare and management of hyperuricemia. Drugs of choice for the acute flare are the nonsteroidal antiinflammatory agents and colchicine. Indomethacin, ibuprofen, naproxen, and sulindac are used initially in maximal recommended doses and are tapered as soon as symptoms abate. Colchicine can be administered either orally or intravenously; it has diagnostic as well as therapeutic value, since no other type of arthritis responds so rapidly and consistently to colchicine.

Hyperuricemia can be managed with a xanthine oxidase inhibitor (allopurinol) or with uricosurics (probenecid or sulfinpyrazone, which increase renal uric acid excretion). Allopurinol inhibits uric acid production by competing with hypoxanthine for the enzyme xanthine oxidase. The solubility and renal clearance of hypoxanthine and other precursor oxypurines are much greater than those of uric acid. Uricosuric drugs act by blocking the tubular reabsorption of MSU. An acute attack of gouty arthritis can be precipitated by allopurinol or the uricosurics, so therapy should begin after the acute reaction and with small doses that are gradually increased.

Calcium Pyrophosphate Dihydrate Disease

Calcium pyrophosphate dihydrate (CPPD) ($Ca_2P_2O_7 \cdot 2H_2O$) can precipitate in ligaments, tendons, articular cartilage, and fibrocartilage. Originally known by the terms chondrocalcinosis or pseudogout, **CPPD crystal deposition disease** can be asymptomatic, can produce a low-grade inflammation, or can present with an acute inflammatory gout-like attack. The mechanism of inflammation is similar to that in gout, with crystal phagocytosis, membrane lysis, and release of lysosomal enzymes into the joint. Inorganic pyrophosphate (PPi) is normally a trace constituent of synovial fluid, and elevated levels most likely come from overproduction by articular chondrocytes.

CPPD crystal deposition disease occurs in about one-half of those patients with hyperparathyroidism or with hemochromatosis. Other associated diseases include diabetes mellitus, gout, hemophilia, and hypophosphatasia, but most cases are idiopathic, affecting middle-aged and elderly men and women (present in 5 percent of 70-year-old people, increasing to 50 percent by age 90. Menisci and articular cartilage of the knees are most commonly involved followed by the triangular fibrocartilage of the wrist, metacarpophalangeal joints, and hips. The symphysis pubis fibrocartilage, the annulus fibrosis of the intervertebral disk, and the glenoid labra are also involved. Radiographs show punctate and linear densities in articular cartilage or in fibrocartilage.

Longstanding CPPD crystal deposition is associated with degenerative joint disease. The articular cartilage becomes fibrillated and eventually erodes with radiographically observable joint space narrowing, bone sclerosis, and occasional subchondral cyst or osteophyte formation. Acute attacks respond to salicylates, NSAIDs, or corticosteroids (in resistant cases), but the response to colchicine is inconsistent.

Basic Calcium Phosphate Disease

Other calcium-containing crystals that can precipitate in and around joints include hydroxyapatite, octacalcium phosphate, and tricalcium phosphate. Because all of these crystals have a basic calcium phosphate component, McCarty proposed the term **BCP crystal deposition disease.** These deposits can induce acute inflammation (acute calcific bursitis/tendonitis), but many patients remain asymptomatic or experience only a mild chronic periarthritis. Some gradually show a destructive arthropathy. The BCP deposits can be intra-articular, in which case they are usually asymmetric and monoarticular; they can also be periarticular, classically involving the supraspinatus tendon or the subacromial bursa. A history of trauma is often but not always present.

Radiographs during the early stages show a thin, cloud-like, poorly defined density. Older deposits are dense, homogeneous radiodensities with sharply defined, irregular borders. Treatment includes physical techniques (heat, ultrasound, exercises) and salicylates or NSAIDs.

HEALING

Wound Healing by First Intention

Figure 9.3 shows a schematic curve of primary wound healing. It is helpful conceptually to divide the healing process into four biological periods: (1) **clot formation**, (2) **inflammation**, (3) **proliferation**, and (4) **remodeling**. Healing of a surgical wound is an example of healing by **first intention (primary healing)**. Those cells in the path of the knife blade are immediately disrupted (including epithelium, fibroblasts, fat, and capillary endo-

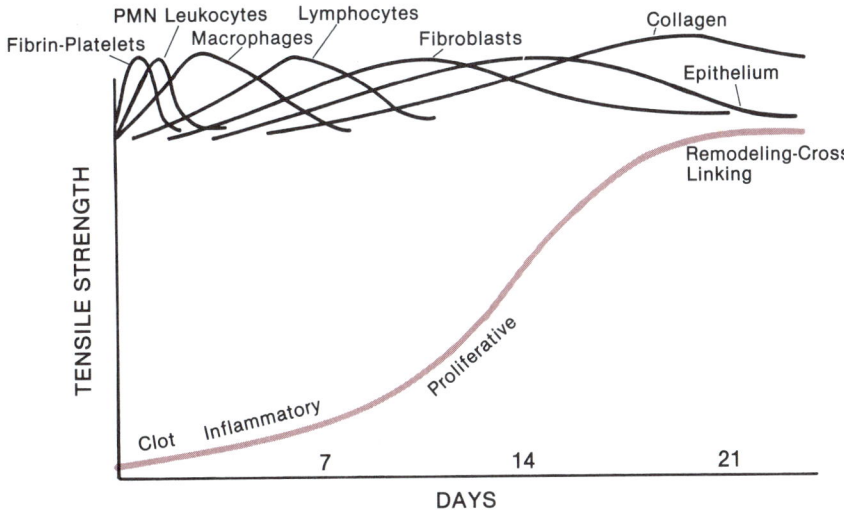

FIG. 9.3. An idealized response curve of wound healing by first intention (primary healing). Arbitrary tensile strength is plotted against time; the four phases of wound healing are noted on the curve. The smaller curves at the top are idealized time courses of the major components of healing. Adapted from Schilling (1976).

thelial cells). Division of capillaries and larger vessels results in bleeding, and fibrinogen, along with other plasma components, floods the area. After approximation of the wound edges by sutures, a thin blood clot fills the space between the wound edges. The surface of the clot dehydrates, forming a scab that seals the wound from the environment. Fibrinogen is the key macromolecule during the first biological period.

Within 6 hours, the second period begins as mediators of inflammation, which were released after injury, attract neutrophils, which engulf any bacteria that may be present. Within 12 hours, the cut edges of epidermis thicken from mitotic activity. Epithelial cells grow and migrate beneath the surface scab, and by 48 hours they meet at the midline, producing a very thin but continuous epidermis. In the absence of any major bacterial contamination, neutrophils rapidly disappear, being replaced by macrophages, which remove damaged and necrotic cells, red blood cells, fibrin, and cellular debris.

By 3 days, healing has progressed to the third period, characterized by proliferation of fibroblasts, capillaries, and epithelium. Collagen is the key macromolecule during the third biological period. New collagen from fibroblasts and new capillary buds (granulation tissue) appear, and by the fifth to the seventh day, granulation tissue fills the entire incisional space. Macrophages complete the removal of debris, and fibroblasts proliferate, secreting collagen fibrils that accumulate and completely bridge the wound. The new collagen fibrils are thin and randomly oriented.

By 10 days to 2 weeks, most of the macrophages have disappeared as the healing process enters the fourth biological period (remodeling). The epidermis has recovered its normal thickness, but the tensile strength of the wound is well below normal. By the end of 4 weeks, connective tissue occupies the entire area with no evidence of the original inflammatory exudate. Gradually, over the next 8 to 12 months, remodeling occurs in which new cross-linked collagen fibers oriented along the directions of skin stress replace the haphazardly arranged collagen. Vascularization decreases, and the area becomes a relatively acellular, hypovascular, pale, collagenous scar. Histologically, the mature scar contains parallel dense bundles of collagen, a few capillaries, and an occasional fibroblast.

Wound Healing by Second Intention

Wounds with a large tissue defect must first be filled prior to epithelialization and healing; this process is known as healing by **second intention (secondary union)**. A large tissue defect has more cellular injury and more necrotic debris than a sutured incisional wound, so the inflammatory reaction is more intense and lasts much longer. Granulation tissue covers a greater surface area and must grow in from the margins of the crater as necrotic debris and exudate are removed.

The major difference between primary and secondary healing is the presence of wound contraction in secondary healing. Much of the defect in large wounds is eliminated by contraction, and the remainder is filled with granulation tissue. Myofibroblasts (modified fibroblasts containing actomyosin) in the granulation tissue and along the wound edges produce the wound contraction (not shortening of collagen fibers). Epithelization and remodeling take much longer in healing by second intention. There is more scar tissue present and greater loss of dermal appendages (hair, sweat glands, etc.).

Some authors write of healing by **third intention,** which is healing with delayed suture closure. A septic or heavily contaminated wound is allowed to remain open while leukocytes and granulation tissue remove bacteria and necrotic debris. Three to 6 days later, depending on the appearance of the granulation tissue, the wound is closed with sutures.

Tendon and Ligament Healing

Similar to wound healing, both ligaments and tendons heal with an initial inflammatory phase followed by fibroplasia, collagen synthesis, and gradual remodeling. However, it is difficult to generalize about tendon and ligament healing because of the structural diversity of tendons and ligaments. For instance, the Achilles tendon, the patellar tendon, and the finger flexor tendons differ in macroscopic and microscopic anatomy, as do the medial collateral ligament, the anterior cruciate ligament, and the deltoid ligament. In any case, after a complete tear or laceration, normal architecture is never completely restored; some scar tissue is always present (of all the mesenchymal tissues, only bone can completely regenerate without a fibrous scar).

The most complex tendon healing concerns finger flexor tendons, which heal from the cut ends and also from the surrounding mesotenon by vascular and fibroblastic ingrowth. The epitenon thickens as fibroblasts begin to divide along the surface and at the cut ends. Macrophages appear within the gap, removing necrotic cells, frayed collagen fibrils, fibrin, and other debris. By 10 days to 2 weeks, the gap is closed by a collagenous "tendon callus." Urbaniak and co-workers showed that the tensile strength of a tendon repair was weakest at day 5 and was not strong enough to withstand mild resistance until 15 to 20 days despite the presence of fibrous repair tissue. Adhesions (fibrous bands tethering the tendon to surrounding soft tissue) are common after tendon healing, but early mobilization reduces their influence on tendon excursion.

The medial collateral ligament of the knee also heals in an orderly process, similar to other soft tissues. Collagen is actively synthesized for up to 6 weeks, followed by gradual remodeling. The difference is that remodeling is extremely slow, commonly taking more than a year, and alterations in collagen content persist, such as a higher proportion of type III collagen in the healed segment (type III collagen forms small fibrils and is deficient in cross links). The anatomy, biochemistry, and biomechanics are never the same as in the normal, noninjured ligaments.

GLOSSARY

B cells Antibody producing lymphocytes that are the precursors of plasma cells.

BCP disease Those diseases that are caused by precipitation of basic calcium phosphate in periarticular tissue including calcific bursitis and hydroxyapatite deposition disease.

Bradykinin An oligopeptide of nine amino acids, released during inflammation, that is the most potent pain-producing chemical known.

Bursa of Fabricius The organ in birds that processes lymphocyte stem cells in such a way that the daughter lymphocytes, B cells, have immunoglobulins on their cell surface and manufacture immunoglobulins for secretion.

Chemotactic factors Compounds that stimulate leukocytes to migrate to an area, escape from capillaries, and accumulate in the area.

Chemotaxis The migration of white cells toward an attractant, usually a mediator of inflammation.

Complement A collection of about 18 serum proteins that are activated in a sequential manner and produce inflammatory effects, membrane lysis, and bacterial killing.

CPPD Calcium pyrophosphate dihydrate.

CPPD deposition disease Also called pseudogout, a joint disease caused by the precipitation of calcium pyrophosphate dihydrate in joints and periarticular tissues.

Cyclooxygenase The rate-limiting enzyme in prostaglandin synthesis, inhibited by aspirin and other NSAIDs.

Cytotoxic T cells Lymphocytes that bind to foreign or virus-infected cells and kill them.

Emigration The process by which white cells escape from blood capillaries into the perivascular tissue.

Exudate An inflammatory extravascular fluid accumulation of plasma, proteins (mainly fibrinogen), and cellular debris.

Fab fragment That part of the antibody molecule containing the hypervariable amino acid sequences responsible for antigen binding.

Fc fragment That part of the antibody molecule that binds to phagocytes and also activates the complement system.

First intention Primary wound healing such as occurs after closure of a surgical wound.

Glucocorticoids Steroids that are powerful inhibitors of the inflammatory reaction and that also stimulate gluconeogenesis. In order of decreasing potency, they are dexamethasone, prednisolone, and hydrocortisone.

Gout An inflammatory disease caused by the precipitation of monosodium urate crystals in joints or soft tissues.

Granulation tissue Admixture of capillaries, endothelial buds, and fibroblasts that appears during the proliferative phase of wound healing.

Helper T cells Lymphocytes of the T-cell line that do not secrete antibody or kill cells but are essential for B-cell antibody response and T-cell cytotoxic effects.

Histocompatibility antigens Cell-surface glycoproteins (HLA antigens) encoded by a group of genes called the major histocompatibility complex located on chromosome 6. Most cells have these antigens, which form the basis of immunological recognition; certain HLA antigens are associated with rheumatic and autoimmune disease.

HLA antigens Cell-surface glycoprotein antigens (histocompatibility antigens) first recognized as human leukocyte-associated antigens.

HLA-B27 A class I histocompatibility antigen associated with seronegative spondyloarthropathies such as ankylosing spondylitis.

Immunoglobulins Protein antibodies synthesized by plasma cells in response to the presence of an antigen. The antibodies specifically react with the antigen that elicited their production.

Inflammation A basic tissue reaction to injury, caused by release of chemical mediators producing vasodilation and extravascular accumulation of plasma and leukocytes.

Leukotrienes Mediators of inflammation derived from plasma membrane fatty acids by the enzyme lipoxygenase.

Macrophage Mononuclear phagocytes, commonly known as histiocytes when found in connective tissue, belonging to the general class of reticuloendothelial cells.

Monoclonal antibodies Immunoglobulins with absolute specificity, isolated in high concentrations either from multiple myeloma plasma cells or from hybrid cells.

NSAIDs Nonsteroidal antiinflammatory drugs such as aspirin, indomethacin, ibuprofen, and other phenylalkanoic acid derivatives.

Pavementing A phenomenon of inflammation in which the endothelium is covered by white cells.

Phagocytosis The recognition, engulfment, and killing or degradation of a bacteria by a white cell, mainly polymorphonuclear neutrophils.

Prostaglandins Five-membered ring structures derived from arachidonic acid in cell membranes, having specific biological activities, including mediating inflammation; PGI_2 and PGE_2 are potent vasodilators.

Second intention Secondary wound healing in which a large tissue defect must be filled in prior to epithelialization; accompanied by wound contracture produced by myofibroblasts in the granulation tissue.

Suppressor T cells Cells responsible for immune tolerance that suppress the response of B cells or T cells.

T cells Lymphocytes that originated from a thymus-processed stem cell and carry out cell-mediated immunity.

Thromboxane Six-membered ring structure found in platelets, derived from plasma membrane fatty acids, produced during inflammation.

Tophi Subcutaneous accumulations in gout consisting of a nodular core of monosodium urate crystals surrounded by a fibrotic capsule.

Uricosuric agents Drugs such as probenecid or sulfinpyrazone that increase renal excretion of uric acid.

Vasoactive substances Compounds released during inflammation such as histamine, serotonin, and bradykinin that relax perivascular smooth muscles and increase blood flow through the injured area.

Xanthine oxidase The key enzyme in uric acid metabolism, responsible for conversion of many purine precursors (usually soluble) into uric acid (which is relatively insoluble). Allopurinol is a xanthine oxidase inhibitor.

BIBLIOGRAPHY

Albers, B., Bray, D., Lewis, J., Raff, M., Roberto, K., and Watson, J. D. (1983): *Molecular Biology of the Cell.* Garland, New York.

Angell, M., and Robbins, S. L. (1979): *Basic Pathology.* W. B. Saunders, Philadelphia.

Bainton, D. (1980): The cells of inflammation: A general view. In: *The Cell Biology of Inflammation,* edited by G. Weissmann, pp. 1–23. Elsevier/North-Holland Biomedical Press, New York.

Frank, C., Schachar, N., and Dittrich, D. (1983): Natural history of healing in the repaired medial collateral ligament. *J. Orthop. Res.,* 1:179–188.

German, D. C., and Holmes, E. W. (1986): Gout and hyperuricemia: Diagnosis and management. *Hosp. Pract.,* 21:119–132.

Hurley, J. V. (1983): *Acute Inflammation,* second ed. Churchill Livingstone, New York.

Manske, P. R., Gelberman, R. H., Vande Berg, J. S., and Lesker, P. A. (1984): Intrinsic flexor-tendon repair. A morphological study *in vitro. J. Bone Joint Surg.,* 66A:385–396.

McCarty, D. J. (1986): Arthropathies associated with calcium-containing crystals. *Hosp. Pract.* 21:109–120.

Moore, T. L., and Weiss, T. D. (1985): Mediators of inflammation. *Semin. Arthritis Rheum.,* 14:247–262.

Resnik, C. S., and Resnick, D. (1983): Crystal deposition disease. *Semin. Arthritis Rheum.,* 12:390–403.

Rodnan, G. P., Schumacher, H. R., and Zvaifler, N. J., eds. (1983): *Primer on the Rheumatic Diseases,* eighth ed. pp. 16–33, 120–133. Arthritis Foundation, Atlanta.

Schilling, J. A. (1976): Wound healing. *Surg. Clin. North Am.,* 56:859–874.

Weissmann, G. (1983): Pathways of arachidonate oxidation to prostaglandins and leukotrienes. *Semin. Arthritis Rheum.,* 13:123–129.

Yunis, E. (1983): The cellular and humoral basis of the immune response. *Semin. Arthritis Rheum.,* 13:89–93.

Subject Index

A
A bands, 127
Action potential, 119–120
Alleles, 7
α chains, 58
Aminoacyl tRNA and synthetases, 6
Amphiarthroses, 101
Antibody (immunoglobulin), 165
Antigens, major histocompatibility, 168–169
antiinflammatory medications, 163–164
Apical ectodermal ridge (AER), 41
Arnold-Chiari deformity, 40
Arthritis
    crystal-induced inflammatory, 169–173
    osteoarthritis, 110–112
    rheumatoid, 112–113
Arthrogryposis multiplex congenita, 42
Articular cartilage, 107–114
    aging of, 109
    diseases of
        chondromalacia, 109–110
        osteoarthritis, 110–112
        rheumatoid arthritis, 112–113
    injury to, 113–114
    origination of, 45
Atlas and axis, embryology of, 38
ATP (adenosine triphosphate), and muscular contraction, 130
ATP synthetase, 24

Autosomal dominant disorders, 11–12
Autosomal recessive disorders, 12
Axial skeleton development, 36–41
Axon(s), 116, 117
    myelinated and non-myelinated, 117
Axonotmesis, 124

B
B cells, 165
B lymphocytes, 165
Base pairs, 2–3
Bioelectricity, 92–93
Blood supply to bone, 86–87
Bone(s)
    cortical and cancellous, 85
        and fine cancellous, 83
    structure of, 85–86
    woven and lamellar, 83
Bone disease, metabolic, 149–156
    marrow packing disorders and malignancy, 156
    osteitis fibrosa, 154–155
    osteomalacia, 153–154
    osteoporosis, 151–153
    renal osteodystrophy, 155–156
    rickets, 153–154
Bone formation, 45–47
    and remodeling in metaphysis, 51–52
Bone grafts, 96–97
Bone growth factors, 90–91

Bone malignancy and marrow packing disorders, 156
Bone mineral, 76–78
Bone morphogenetic protein, 77
Bone morphology and biology, 81–99
　basic physiological processes, 87–92
　　growth, 87–88
　　modeling, 88
　　remodeling, 88–89
　bioelectricity and, 92–93
　blood supply and, 86–87
　bone as material, a tissue, an organ, 81–82
　fracture healing and, 93–96
　myositis ossificans and, 97
Bone surfaces and membranes, 84–85
Bone tissue, classification of, 82–84
Bradykinin, 163
Brown tumors, 155
Brushite, 76
Bursa of Fabricius, 165

C

Calcification, 78
　abnormal, 156–158
　provisional, 50
Calcitonin, 149
Calcitriol, 143–144
Calcium and phosphorus
　biochemistry of, 138–141
　and hormonal regulation of bones, 143–149
　and metabolic bone disease, 149–156
　physiology of, 141–143
Calcium binding protein, 144–145
Calcium phosphate disease, basic, 172–173
Calcium pyrophosphate dihydrate (CPPD) disease, 172

Carrier-mediated active transport, 19
Cartilage, articular, see Articular cartilage
Cartilage varieties, 106–107
Catabolin, 113
Cathepsin series, 65
Cell
　typical, 16–18
　undifferentiated mesenchymal, 25–26
Cell-mediated immunity, 167–168
Cells and subcellular organelles, 16–31
　chondroblasts and chondrocytes, 26–27
　fibroblast, 29–30
　osteoblasts and osteocytes, 27–28
　osteoclast, 28–29
　plasma membrane, 16, 18–20
　subcellular organization, 20–25
　　cytoskeleton, 25
　　cytosol, 20–21
　　endoplasmic reticulum, 21–22
　　Golgi apparatus, 22–23
　　lysosomes and peroxisomes, 24–25
　　mitochondrion, 23–24
　　nucleus, 20
Centromere, 7–9
Chemiosmosis, 24
Chemotactic factors, 161
Cholecalciferol, 143
Chondroblasts, 26–27
Chondrocytes, 26–27
Chondromalacia, 109–110
Chondronectin, 75
Chromatids, 7
Chromatin, 9
Chromatolysis, 124
Chromosomal disorders, 12–13
Chromosomes, anatomy of, 7–10
Clear zone, 29

SUBJECT INDEX

Cloning, 1
Codons, 4
Collagen(s), 56–67
  in basal lamina and cytoskeleton, 60
  interstitial (fibrillar), 60
  molecular heterogeneity of, 59
  structure of, 58–59
  types and properties of, 59
Collagen metabolism, disorders of, 65–67
Collagen synthesis, 60–62, 63
  posttranslational modifications and fibrillogenesis and, 63, 64–65
  structure and function of procollagen and, 62
Collagenases, 65
Complement, 167
Congenital malformations, 13
  genetic or environmental factors in, 48
Contractions, isometric versus isotonic, 133
Creatine kinase, 130–131
"Creeping substitution", 95
Cross linking, 64–65
Crystal growth, 77
Crystal-induced inflammatory arthritis, 169–173
Crystal nucleation, 77
Cyclooxygenase, 163
Cylinderization, 54
Cytoskeleton, 17, 25
Cytosol, 16–17, 20–21

D

Dendrites, 116
Deoxyribonucleic acid (DNA), 1–5
  in chromosomes, orders of packing of, 10
  conversion of information in (to mRNA to protein), 4–7
  genetic code of, 4
  structure of, 2–4
Dermatome, 34
Desmosine, 73–74, 75
Diaphyseal cortex of long bone, 84
Diarthroses, 101
1,25-dihydroxycholecalciferol, 143–144
Diploid cell, 7
DNA, see Deoxyribonucleic acid
Dupuytren contracture, 67

E

Ectoderm, 33
γ Efferents, 123
Ehlers-Danlos syndrome (EDS), 66
Elastase, 74
Elastic cartilage, 107
Elastin, 73–74
Embryology and growth, 33–55
  axial skeleton development, 36–41
  bone formation, 45–47
  embryonic period, 35–36
  fetal period, 47–48
  functional anatomy of growth plate, 48–53
  circumferential structure, 52–53
  metaphysis, 51–52
  morphology of growth cartilage, 49–51
  further growth of bone (modeling), 53–54
  germ cell layers: endoderm, ectoderm, mesoderm, 33–35
  joint, 44–45
  limb, 41–44
Embryonic period, 35–36
Encephalitis, allergic, 117
Endochondral bone formation, 46
Endocytosis, 19
Endoderm, 33

SUBJECT INDEX

Endomysium, 126
Endoneurium, 121
Endoplasmic reticulum, 17, 21–22
Enthesis, 101
Epimysium, 126
Epineurium, 121
Epiphyseal artery, 49
Epiphysis and metaphysis and diaphysis, 53
Excitation-contraction coupling, 129
Exercise
    isokinetic, 134
    isometric or isotonic, 133
Exons, 5, 6
Extracellular matrix, 57

F

Fc fragment, 165
Fetal period, 47–48
α-fetoprotein, 39
Fibrillogenesis, 62, 63
    and posttranslational modifications, 63, 64–65
Fibroblast, 29–30
Fibrocartilage, 107
Fibronectins, 74
First intention (primary healing), 173–174
Fracture healing, 93–96
    cancellous, 95
    external bridging callus, 93–95
    nonunion and, 96
    primary bone, 96
Free diffusion, 19
Funnelization, 53

G

Gene expression, regulation of, 7, 8
Genetic code, 4, 5
Genetic diseases, three types of, 11–13
Genetics
    molecular biology and, 1–15; see also Deoxyribonucleic acid
    anatomy of chromosomes, 7–10
    regulation of gene expression, 7–8
    transcription (DNA to RNA), 4–5
    translation (RNA to protein), 5–7
    of some musculoskeletal diseases, 10–13
    multifactorial disorders, 13
Germ cell layers: endoderm, ectoderm, mesoderm, 33–35
Glucocorticoid steroids, 164
Glycine, 58
Glycocalyx, 19–20
Glycosaminoglycans (GAGs), 68–69
    GAG side chains, 71
Golgi apparatus, 17, 22–23
Gout, 170–172
Grafts, bone, 96–97
Growth, see Embryology and growth

H

H zones, 127
Haversian system (osteone), 86
Healing, 173–176
Hemispherization, 54
Hensen's node, 34
Hilton's law, 102
Histologic zones, 48
Histones, 9
HLA antigens, 168
Hormonal regulation of bones, 143–149
    calcitonin in, 149
    parathyroid hormone in, 146, 148–149
    vitamin D in, 143–146, 147
Howship's lacunae, 28
Humoral immunity, 165–167
Hurler disease, 25
Hyaline cartilage, 106

SUBJECT INDEX

Hyaluronic acid, 68
Hydroxyapatite, 76
Hydroxyproline, 58–59
Hyperparathyroidism, primary, 148
Hypoparathyroidism, 148

I
I bands, 127
Immunoglobulins, 165–167
Immunology, 164–169
  cell-mediated, 167–168
  humoral, 165–167
  major histocompatibility antigens, and, 168–169
Inflammation, 161–164
  antiinflammatory medications and, 163–164
  cardinal signs of, 161
  chronic, 164
  and healing, 173–176
  and immunology, 164–169
  mediators of, 161–163
Inflammatory diseases, crystal-induced, 169–173
  basic calcium phosphate (BCP crystal deposition) disease, 172–173
  CPPD (crystal deposition disease), 172
  gout, 170–172
Inflammatory reaction, 162
Interzone, 44
Intima, 102
Introns, 5, 6
Isometric or isotonic exercise, 133

J
Joint morphology, 100–101
Joints, 44–45, 100–115
  articular cartilage and, 107–114
  cartilage varieties and, 106–107
  embryology of, 44–45
  synovial fluid and, 103–106
  synovium and, 101–103

K
Klippel-Feil syndrome, 40–41
Krebs citric acid cycle, 17–18
Kyphosis, congenital, 39

L
Lamellae, Haversian or non-Haversian, 83, 84
Lathyrism, 65
Leukotrienes, 163
Ligament and tendon healing, 175–176
Limbs, embryology of, 41–44
Link proteins, 71
Lipoxygenase, 163
Looped domains, 9
Looser's zones, 154
Lubricin, 103
Lymphoid cell line, 28
Lysosomes, 17, 24–25

M
Major histocompatibility complex (MHC), 168
Marfan syndrome, 66
Marrow packing disorders and malignancy, 156
Matrix modification, 77
Membrane potential, 118–119
Mendelian disorders, 11–13
Mesenchymal cell, undifferentiated, 25–26
Mesoderm, 34
Metabolic bone disease, 149–156
Metaphysis, 48
  bone formation and remodeling in, 51–52
Mitochondrion, 23–24
Modeling, 53–54, 88
Molecular biology, *see* Genetics
Monoclonal antibodies, 167
Motor unit, 120
Mucin clot, 104

Mucopolysaccharidoses, proteoglycan synthesis degradation and, 72–73
Muscle, nerve and, see Nerve and muscle
Muscle atrophy, 134
Muscle cytoplasm, 126
Muscle fibers, Type I and Type II, 131–132
Muscle structure, 125–126
Muscle tone, 123–124
Musculoskeletal diseases, genetics of some, 10–13
Mutations, 4
Myelomeningocele, 38–39
Myofibroblast, 67
Myoglobin, 126
Myositis ossificans, 97
Myotome, 34

N

Nerve and muscle, 116–137
  action potential and, 119–120
  contractile properties and, 132–133
  effects of training and, 133–134
  energy metabolism and, 130–131
  excitation-contraction coupling and, 129–130
  membrane potential and, 118–119
  muscle fiber type and, 131–132
  muscle structure and, 125–126
  muscle ultrastructure and, 126–129
  peripheral nerves, 121–125
  synapses, 120–121
Nerve injury, mechanical, 124
Neuroma, 124
Neuron, 116–118
Neuropathies, 124–125
Neuropraxia, 124
Neurotmesis, 124
Neurotransmitters, excitatory and inhibitory, 120
Neurovascular hiatus, 126
Nodes of Ranvier, 117
Nonsteroidal antiinflammatory drugs (NSAIDs), 163
Notocord, 34
Nuclear envelope, 20
Nuclear lamina, 20
Nucleolus, 9, 20
Nucleus, 20
Nutrient artery, 49

O

Ossification, 78
Ossification groove of Ranvier, 48, 52–53
Osteitis fibrosa, 154–155
Osteoarthritis, 110–112
Osteoblasts, 27–28
Osteocalcin, 77
Osteoclast, 28–29
Osteoconduction, 96–97
Osteocytes, osteoblasts and, 27–28
"Osteocytic osteolysis", 148
Osteogenesis imperfecta (OI), 66–67
Osteoid, 76–77
Osteoinduction, 97
Osteomalacia, 153–154
  in chronic renal failure, 155
  vitamin D deficiency or resistance causing, 145
Osteonectin, 77
Osteones, 86
Osteopenia, 149
Osteoporosis, 151–153
Oxidative phosphorylation, 18

P

Paget's disease, 89
Pannus, 113
Parathyroid hormone, 146, 148–149
Perimysium, 126
Perineurium, 121
Perioxisomes, 17, 25

Perichondrial arteries, 49
Perichondrial ring of LaCroix, 48, 52
Peripheral nerves, 121–125
  muscle tone and, 123–124
  mechanical injury to, 124
  neuropathies and, 124–125
Phosphocreatine, 130
Phosphoproteins, 77
Phosphorus and calcium, biochemistry of, 138–141
Plasma membrane, 16, 18–20
Podagra, 170
"Prepro-PTH", 146
"Pro-PTH", 146
Procollagen, structure and function of, 62
Progenitor cells, 25–26
Proline, 58
Proprioceptors, 122
Prostaglandins (PGs), 163
Protein, DNA and RNA in, 4, 5–7
Proteoglycan subunits and aggregates, 69–72
Proteoglycans, 67–73
  glycosaminoglycans (GAGs), 68–69
  synthesis and degradation of, 72–73
  versus tissue glycoproteins, 72
Pseudofractures, 154
Pseudohypoparathyroidism, 148–149

R
Reflex arc, 122
Remodeling, 77, 88–89
  of hard callus, 95
Renal osteodystrophy, 155–156
Rheumatoid arthritis, 112–113
Rheumatoid factors, 112
Ribonucleic acid (RNA), 3–7
  aminoacyl tRNA and synthetases, 6
  DNA goes to (transcription), 4–5
  goes to protein (translation), 5–7
  heteronuclear (hnRNA), 5
  messenger (mRNA) and ribosomal and transfer (tRNA), 4
  polymerase, 4
Ribosome(s), 6, 21–22
Rickets, 153–154
  or osteomalacia, vitamin D deficiency or resistance causing, 145
  type I vitamin-D-resistant, 146
RNA, *see* Ribonucleic acid, 3–7
Ruffled border, 29
"Rugger jersey spine", 156

S
Saltatory conduction, 120
Sarcomere structure and function, 127, 128–129
Sclerotome, 34
Sclerotomic resegmentation, 38
Second intention (secondary union), 174–175
Segmentation, defects of, 39–40
Sliding filament model, 128
Somites, 34
Spina bifida occulta, 38
Spinal formation, 37–41
Spine, congenital deformities of, 38–41
Spongiosa, primary and secondary, 52
Stress-generated potentials, 92
Stretch reflex, 123
Structural components of musculoskeletal system, 57–79
  collagen, 57–67
  elastin and fibronectin and other proteins, 73–76
  extracellular matrix, 57
  mineral, 76–78
  proteoglycans, 67–73
Subcellular organelles and cells, *see* Cells and subcellular organelles

Subcellular organization, 20–25
Subintima, 102
Synapses, 116, 117, 120–121
Synarthroses, 100–101
Synovial fluid, 103–106
  of non-Newtonian properties, 103
  Synovial fluid analysis, 104–106
Synovial plica, 45
Synovial pseudarthrosis, 96
Synovium, 101–103

T
T cells, 167–168
Tay-Sachs disease, 25
Tendon and ligament healing, 175–176
Thick and thin filaments, 128
Thromboxane A_2, 163
Tidemark, 108
Tophi, 170
Transcription, process of, 4–5
Translation, process of, 5–7
Triplets, 4, 5

Tropocollagen, 58
  versus proteoglycans, 72
Tropoelastin, 73

V
Vasoactive substances, 161–162
Vitamin D, 143–146, 147
Vitamin D deficiency or resistance causing rickets or osteomalacia, 145
Volkmann's canals, 86

W
Wallerian degeneration, 124
Wolff's law, 88
Wound healing by first or second intention, 173–175

X
X-linked recessive diseases, 12

Z
Z lines, 127

A113 0932868 7

SCI QP 301 .G35 1987

Gamble, James Gibson.

The musculoskeletal system